Cover Art by Tia Lowenthal

Copyright © 2021 by Linda McIver

All rights reserved. No part of this book may be reproduced in any manner whatsoever without written permission except in the case of brief quotations embodied in critical articles and reviews.

First Printing, 2021

Dr Linda McIver pioneered authentic Data Science and Computational Science education with real impact for secondary students and founded the Australian Data Science Education Institute in 2018. Author of Raising Heretics: Teaching Kids to Change the World, Linda is an inspiring keynote speaker who has appeared on the ABC's panel program Q&A, and regularly delivers engaging Professional Development for Primary, Secondary, and Tertiary Educators across all disciplines.

A passionate educator, researcher and advocate for STEM, equity and inclusion, with a PhD in Computer Science Education and extensive teaching experience, Linda's mission is to ensure that all Australian students have the opportunity to learn STEM and Data Science skills in the context of projects that empower them to solve problems and make a positive difference to the world.

Raising Heretics

TEACHING KIDS TO CHANGE THE WORLD

Dr Linda McIver

Australian Data Science Education Institute

Raising Heretics is dedicated to

The extraordinary Jacky Pallas, a born heretic if ever there was one,

Andrew, who encourages my heretical nature with surprisingly little fear of the consequences,

And Zoe and Solara, who are so much more enlightened, aware, and extraordinary than I have ever been. Our future is in excellent hands.

Contents

About The Author — iv
Foreword — ix
Introduction: Why Heresy? — xi

1. Who's in charge? — 1
2. The shape of the future — 28
3. Science is Solved — 58
4. Measurable or Meaningful: Pick One — 77
5. Accepting the Unexpected — 103
6. Projects with Impact — 122
7. Templates for Data Science Projects — 140
8. What Now? — 166
9. Sample Projects — 171
10. Acknowledgements — 183

References and Endnotes — 187

Foreword

My whole career I've been trying to figure out how to solve the problem of teaching programming. I started with teaching first year Computer Science students, and running extension classes with Primary School kids. Then I worked with year 10s and year 11s, before teaching teachers, and even teaching people how to teach teachers. And out of all of that experience, the one thing I know for sure is that motivation is everything, and the strongest motivation comes from the power to solve real problems. "Computing with purpose", my friend Dr Nicky Ringland calls it.

But the more I worked with these motivated kids, and then with motivated teachers, the more obvious it was that teaching programming was only a small part of the importance of doing these real projects. This isn't just about programming, or even just about STEM. This is about saving the world. About building a future that works for all of us, where we can all be healthy, creative, and safe.

This book contains my hope for the future. My dream of an education system that empowers kids to change the world. But that hope, and those dreams, were born in my classes, and in my home. As I watch my own two teenagers fiercely tackle injustice, and I see my former students battle to make the world a better place, I realise that these kids have the power to save us all. And we have the power to help them do it.

Please engage with the ideas in this book. Explore them. Challenge them. Test them. And join the conversation online, by joining the Teachers Using Data Science Facebook group, or engaging with @DataSciAu on Twitter. And if you like what you read here, support The Australian Data Science Education Institute's work at https://www.givenow.com.au/adsei

Many of the footnotes in this book are in the form of website addresses. To save you laboriously typing in every URL, you can find a full list of clickable footnotes from the whole book at https://adsei.org/raising-heretics-footnotes/

If you like the book please share it, share the ideas, and make some noise. And go raise some heretics!

Linda McIver

June 2021

Introduction: Why Heresy?

It's time to change the world. We need creative problem solvers to address catastrophic climate change, income inequality, pandemics, ecological collapse, misinformation, radicalisation, and many more problems facing humanity. We need critical thinkers. Rational Sceptics. People willing to challenge the status quo.

Unfortunately, we have an education system that's exceptionally good at turning out obedient people full of "facts" and unshakeable opinions. This book proposes a new approach to education that empowers our children to solve real problems and , to challenge their own results, and to shake up the status quo on the basis of evidence and data.

I founded the Australian Data Science Education Institute in 2018 because I wanted to show kids that they are capable of working with technology, that it wais relevant to them, and that they don't have to look like Sheldon from the Big Bang Theory in order to learn to program.

It's well known that the technology industry has a diversity problem when it comes to women, but lack of diversity goes way beyond gender. By trying to increase the number of women and girls in STEM, we are only tackling the easy part – though it's actually not that easy, judging by the sheer volume of women in STEM programmes and the persistently stubborn failure of the numbers to actually shift.

The problem is that we consistently attract the kinds of people to tech that are already there. We are missing big chunks of the population – boys included. Boys who don't see themselves as nerdy, or who don't see the point of tech. Girls who don't see it as relevant to them. Non binary and gender queer kids who don't see themselves as represented or welcome in any of the tech programmes available to them.

If we had true diversity in technology and Data Science, we'd have a range of ethnic and cultural backgrounds, as well as people with a wide range of physical abilities. We'd have people on our design teams that are mobility compromised, vision impaired, with allergies, with varied gender identities and sexualities, with every possible skin tone and body shape. We'd have people who act differently, dress differently, think differently, and have different needs. I have headphones that don't work well with long hair, for goodness' sake! Guess who was on that design team?

This lack of diversity is bad for the technology industry, but it's even worse for the rest of us, because technology is changing the shape of our world at an alarming rate, and we currently have very little say in our own future. Companies like Uber and Doordash are radically changing our working conditions and eliminating hard won entitlements and protections[1], while Facebook and Youtube spread misinformation and encourage radicalisation[2], all in the name of keeping people on their platforms and maximising their profits. Our world is being directly shaped by technology companies that are working in ways we don't understand and have no control over.

Meanwhile we see human resources companies using AI to filter job applicants, claiming that their system eliminates "human bias"[3], without admitting the possibility that it introduces new forms of machine bias. We see "predictive policing" algorithms being used to predict crime and target particular communities in disturbing ways[4]. We see a rush towards machine learning and artificial intelligence systems for their own sake, rather than for problems they can legitimately solve, and we have a wholly unwarranted confidence in the accuracy, reliability, and objectivity of their output.

It turns out that diversity in the technology industry is only a small part of the reason why teaching all kids Data Science and STEM skills matters. The big part is that we need a technology and data literate population who are trained to think critically and creatively, and, in particular, trained to believe that they can solve problems. That's the world we need to build. And the foundation stone of world building has to be education.

We have a choice. We can train kids to be obedient process followers who don't rock the boat, or we can train them to be challenging, critical and creative thinkers who ask difficult questions and come up with innovative solutions to our worst problems.

Above all, we need people who are prepared to be heretical.

Who ask "why?"

Who ask "how can we be sure?"

Who ask "what have we missed?"

Who ask "how can we do better?"

Who ask "who are we hurting?"

Who ask "how can we fix this for *everyone?*"

Who ask "how will we know how well it works?"

These questions are often heretical. By asking them, I've sometimes made my bosses very unhappy. They make people uncomfortable. But they are crucial to building an ethical, sustainable, positive future for all of us.

I have a PhD in Computer Science Education and over twenty years experience teaching Computational and Data Science at both Secondary and Tertiary levels. Now I'm the Founder and Executive Director of the Australian Data Science Education Institute (ADSEI) – a registered charity dedicated to ensuring every student is empowered with data literacy, Data Science, and STEM skills. I started ADSEI because I figured out how to engage kids with STEM and Data Science skills, and I wanted to engage *all* kids, not just the kids in my own classes. I thought this would help improve diversity in the technology industry, but I have come to realise the problem is far more fundamental than that.

All of my time in education has made it clear to me just how badly wrong education has gone. We continue to make the same educational mistakes we've been making for decades. We are failing our children, and, in doing so, we are sabotaging our future. If we want to build a future that is evidence based, rational, and inclusive, then our education system clearly needs to change.

There are so many signs that our current education system is missing the mark. When my teenager gets frustrated because she doesn't understand how what she's learning in maths could ever be useful. When a primary school kid says science is boring. When a high school kid says maths is too hard, or science isn't for them, or they aren't smart enough to program a computer. None of these things would happen if education was working. It's obvious that it's not.

And that's unsurprising, since the primary focus of education is a matter of facts, rote learning, and mindless application of procedures. By giving kids "experiments" to do that have known inputs and known results, we teach science as confirmation bias. This trains them that the important thing is to get the right, expected answer (and if you get a different answer, fudge things until it's right!), rather than exploring the unknown and looking for new things.

Although the importance of STEM is widely acknowledged, it is frequently taught as a matter of tech toys, rather than a crucial tool for solving real problems. This commonly comprises a day of robotics play, or the installation of a maker space where kids can tinker with 3D printers and laser cutters. These toys are frequently error prone and difficult to use, so when kids don't find them fun, or have trouble using them, they assume that STEM is something they can't do.

Even when problem solving tools like Design Thinking are introduced in the classroom, they are often only used to solve toy problems that don't relate to challenges that kids can tackle in real life. Design Thinking plays with trips to Mars, or responding to a famine in Ethiopia, instead of taking one of the many problems in our own schools and communities and empowering kids to solve it. You can't teach problem solving properly if you skip the really tough part; implementing your solution and then troubleshooting all the ways it doesn't work the way you thought it would.

By doing this, we tell kids that they can't make a difference until they are grown up, when we could be giving them the tools to make a positive difference in their world today.

The truth is, with this kind of education we have got really good at turning out obedient kids who follow the rules and do as they are told. And those are not the kind of people we need to overcome the huge crises we're facing. We need people who are confident, skilled, knowledgeable, and prepared to stand their ground and argue a point. We need people who see things differently, who look for new answers, who understand uncertainty, and who ask hard questions. We need people who are "unbossable", who don't do what they're told without first understanding why it's the right thing to do. We need people who challenge the status quo. We need people who consider ethics first, rather than as an afterthought or not at all.

Meanwhile, Science has somehow become a partisan political football. Australia's response to the Covid19 crisis was effective, largely because the State Governments followed the advice of experts in epidemiology. Unfortunately, we face a larger and more serious existential crisis in the form of climate change, and in this case, the Government is ignoring experts and investing deeply in denialism and cheap grabs for immediate power and profit.

Policy in this country (and most of the world) is largely driven by ideology, powerful lobby groups, and manipulative media organisations, rather than by science and evidence. This kind of destructive behaviour is justified with dodgy data and deeply suspect visualisations, and all too often even the media lack either the scepticism or the skill to call them out.

Inequality is rising under the influence of capitalism-driven globalisation that promises better lives for all via the concept of "trickle down economics", which the data shows quite clearly does not work. We resist Universal Basic Income on the basis that people would stop working out of laziness, when the data from the trials so far shows not only that people don't stop working, but also that they become more entrepreneurial[5]. Our governments sell off natural assets, log native forests, privatise essential services like health and education, and give tax cuts to big business[6] despite evidence showing that the best way to stimulate the economy is to give money to poor people. As a population, we swallow the line that it is all for our own benefit, and vote the same people back in.

Social media also drags us by the nose, constructing ever more cunning ways to tie us to their platforms, milk us for data and profit, and manipulate our behaviour, all without our informed consent. Our social and workplace gains are casually undermined by disruptive technologies, while we have no input into, and even less control over, the way they shape our future.

This is why we need a rationally sceptical population. We need to stop being irrationally sceptical of climate science and vaccines and start being rationally sceptical of government policy, business motives, and media beatups.

Why Data Science?

How can Data Science Education get us there?

When I was teaching Computational and Data Science at a Science School in Victoria, I taught two subjects. One was a year 11 subject in which I had complete freedom to explore my idea of a broad and interesting Computational Science subject. My teaching partner, Victor Rajewski, and I built a subject that included Artificial Intelligence, Data Science, and Usability, as well as Python programming, Data Structures, and Algorithms. It was essentially an overview of a range of Computer Science topics.

The capstone project for that subject was a Computational Science assignment where students worked with scientists to solve their data and computational needs. The scientists we worked with knew that, like any student project, there was no guarantee they would get any software out of it that they could use; but the students knew that they were working on a real problem that could make a material difference to scientific research if they achieved their goals.

That first class of year 11s worked on cancer and marine biology research for their capstone project. Some groups were content to complete the assignment and move on, but in that first year one group did actually produce a system that made a material difference to a cancer researcher's work. They were so enthusiastic about the project that they continued working on it well into the next year, refining and improving the efficiency of the program and adding more features. In year 12, which many people say is the hardest year of your life (they lie, but that's another story), these two students continued to work on something for which they could no longer get subject credit, simply because it was *worth doing*.

Incidentally my faculty leader at the time had taken one of those two students, Chris, aside at the start of the year and suggested he work for a software firm rather than do my subject, because he felt there was nothing we could teach a student who was already so advanced in programming. Like many teachers, he could not envisage a way of engaging an advanced student with the standard curriculum.

That was a real turning point for me. I confess that in that first year, my first experience of secondary teaching, many of the other assignments we did earlier in the course were far less inspiring. But when we gave the kids something *real* to do, with meaning and purpose, they absolutely flew. Not all of the projects went on to be used in research, of course. But all of them were meaningful. The kids got to work with real data in all its messy, rich, complex, and downright difficult glory, on authentic problems that had the potential for real world impact.

In the years that followed we worked on Neuroscience, Psychology, Astrophysics, Microbiology, and Conservation Ecology projects, among others. In subsequent years we have repeatedly had kids continue with their projects long after the assignment was over, sometimes even into the following years, and we even had a student who wasn't enrolled in the subject beg me to be allowed to do the assignment. There can be no doubt that we succeeded in creating engaging, motivating projects.

That year 11 course was an elective, and in the first year we only had around 30 students. Was it just an easy win, with particularly motivated students who already wanted to study computation? Let me tell you, then, about the subject aimed at 15 or 16 year olds called Creative Studies, which was compulsory for every student in year 10 – around 200 of them.

That subject initially endeavoured to teach Computing the "fun" way. We did simulation in a system called StarLogo, which was a kind of precursor to Scratch, essentially a block-based programming environment designed to attract children. We did robotics with LEGO Mindstorms NXT2 robots, where they had to program the robots to push each other out of a circle, or follow lines. We did experiments with slime mould following mazes to explore natural computation. And we did a small amount of Python programming, where the kids wrote small programs to play simple games like 21 and tic tac toe.

The kids hated it. They were passionate science students who were going to need computation as a crucial foundation to their future careers, yet they could not see the point. They saw it as time wasted playing with toys. Many of them wound up cheating to get through assignments that they found too difficult and too meaningless to waste time on. We persisted with this approach for several years, with a range of variations, always trying to make Computer Science fun. What we succeeded in doing, for the most part, was in making Computer Science frustrating and meaningless.

It was frustrating because the systems we were using were toy programs designed for younger kids, so the students who wanted to do more complex and interesting things, especially in StarLogo, kept bumping up against the limitations of those systems.

It was meaningless because they couldn't see any point to all of this effort that they were pouring into "toys" that they simply didn't find fun.

And it was frustrating because the kids who weren't already into programming, or who thought they couldn't do it, were stumped when they hit a problem with their code. They had teachers who, for the most part, didn't know how to fix their problems, and they were programming in systems that were designed for hacking around and having fun, not debugging. It was very difficult to make what was going on visible so that they could understand their errors. Usability guru Jakob Nielsen calls this "visibility of system status"[7], and it is crucial to learners, or indeed any users, being able to build up any idea of how the system works.

When we switched to more modern block based programming systems a year or two later, based on MIT's Scratch programming language, I once spent 15 minutes trying to debug an issue with a student's code that turned out to be a divide by zero error. Rather than issue a meaningful error message, or even display the offending values, the script simply silently stopped, leaving the student with no idea what was going on. Had there been an error message, as there would be in most programming languages, the debugging would have been almost instantaneous.

I used a careful, systematic trace through of the code, and it still took me ages to find it. A teacher with less programming experience would have very little chance of ever understanding the problem, which would leave the student with code that didn't work, and no-one who could tell them how to fix it. Indeed, that student was sent to me by their teacher, who couldn't solve the problem.

Not the most motivating or inspiring experience of Computer Science! These kinds of experiences teach kids that Computer Science is something they simply can't do. It also taught them that I was a genius who could fix things nobody else could. So it was great for *my* ego, but really bad for their own. The kids loved me, but hated my discipline. Not what we were trying to achieve! It was hands down the most hated subject in the school, even overtaking Physical Education.

Eventually I managed to persuade the teacher in charge to try teaching the same basic programming skills in the context of Data Science. We used real datasets. In the first version of the course we used some data downloaded from the Australian Electoral Commission (AEC). We taught the same fundamental programming skills, but now we used a real world programming language, Python, and a problem where the kids could find meaning and purpose.

We also taught data literacy skills, including figuring out what questions the data could answer, how to create a compelling and valid visualisation (often hand drawn), and what to do with errors and outliers in the data. Kids had to find their own question that the dataset could answer, so they weren't able to cheat, because every assignment was different.

It turns out that "what questions can this dataset answer?" is a much trickier question than you might think. With the AEC data we had a file containing every vote cast in Victoria for the senate in a federal election. The first question the kids said it could answer was "which is the best party?" The trouble with the term "best" is that it's entirely subjective. So this led to a wonderfully rich conversation about the difference between qualitative questions, such as which is the "best", "nicest", or "worst", and quantitative questions, such as which is the "fastest", "longest", or "most successful" (for a specific definition of "successful"). Data can often answer quantitative questions, but it can't answer qualitative ones.

Who got the most votes? Sure!

Who is the best? What do you mean by "best"?

As we refined the subject over the next few years, we used different data, including climate change projection datasets and microbat research data, and the interesting thing was that the students didn't seem particularly bothered by *which* real dataset we used, or what topic we chose.

The fact that the dataset was real, and that the questions we asked about the data were purposeful, meant that the kids were motivated. They understood, for the first time, why we were teaching them this stuff. Those kids aren't all going to become data scientists, but they all acquired technical skills they might not have thought were worth learning before, and understood the value of those skills. They also have data literacy that they can use, not only when handling their own data, but when critically evaluating data from other sources – like politicians, and the media!

Suddenly the subject had a purpose. Even though we still did not have enough skilled programming teachers to help the students debug their code, and there were many other problems with the first couple of times we ran the subject, the students now had the motivation to persist. They could see the point. The kind of kids who used to say in feedback surveys:

"Why are you making me do this?",

"It's not relevant to me.",

"I can't do it.",

"I'm not interested.";

were now saying

"this is so important",

"this is so useful";

and even

"I'm using this in my other work".

Plus my all time favourite:

"I did something I didn't think I could do!"

Finally we had found what drives kids to learn technical skills. It's not for fun. It's not even for marks. It's for meaning. Give kids something meaningful to do, and they will smash it out of the park, and even engage with it beyond the scope of the course.

I want to tell you another story that illustrates the difference between the two subjects quite effectively. Back when the year 10 subject was still aiming for "fun", I had a student in my year 10 class named Austin. Austin quickly became my nemesis. His definition of fun was very different to ours, and he could not see the point of the hoops we wanted him to jump through *at all*. Austin was bored. He was sullen. He was disruptive. He didn't do a stitch of work all year.

To be honest, I had a certain sneaking sympathy for him, because I hated that subject too. I was pushing hard to get it changed, but we both had to deal with it the way it was, and he made it *hard*. I pushed him. He pushed back. We achieved nothing. I think we both heaved a great sigh of relief when the year came to an end.

And then, when I was checking my class list for the next year's crop of year 11s, there he was. I couldn't believe it. My family can attest that I ranted and raved. "Why would he do this to me?" (Of course, it was all about *me!*) "Why can't he go do some other subject? Do I really have to go through another year of this???"

The first class we had together in the new subject, I taught them all some basic Python skills and started them working on a rudimentary chatbot – a computer program that pretends to be a human, and chats with the user. Always one to do things differently, Austin came to school *the next day* with 1000 lines of Python code (having never programmed in Python, or anything like it, before) that was a near perfect English to Pig-Latin translator.

In case you are unfamiliar with Pig-latin, the rules are simple. For words that start with a consonant, take the first letter of the word, put it on the end and add "ay". So "take" becomes "aketay". "Word" becomes "ordway". "Consonant" becomes "onsonantcay". Words that start with a vowel simply add "ay" on the end, and words that start with more than one consonant shift both consonants to the end. So "start" becomes "artstay".

I broke it on 'Qantas' (because the program assumed any word starting with 'q' would have 'u' as its next letter, though using Qantas is cheating, because Qantas is an acronym, not a real word) and 'rhythm' (which is not cheating, because rhythm IS a real world, but it is most unusual for an English word that starts with 'r' to have 'h' as the next letter). The next day he came back with those two examples fixed, and the code cleaned up. He went on to modularise it, comment it, and polish it. I was astounded. Where had he been the previous year??? But, in truth, I knew where he had been. He had been unmotivated, uninspired, and uninterested. And who could blame him?

Austin went on to be one of the best students I ever had. Even his problem specification documents, the parts few serious coders enjoy, were polished and professional. He worked in a team to produce an outstanding Computational Science project to monitor dolphin populations in partnership with Polperro Dolphin Swims[8]. His determination, work ethic, and drive, which were completely absent the previous year as far as I could tell, were phenomenal. Austin is my cautionary tale (and I tell his story with permission – we are now good friends). He is a perfect example of the difference between engaging kids by giving them something meaningful to work on, and trying to get them interested by giving them "fun toys" that have no purpose.

I was having a wonderful time in that classroom, and I could have stayed there very happily forever. But it wasn't enough for me to give just the kids in my classes the opportunity to engage with tech skills and build their data literacy with authentic, meaningful projects.

I wanted all kids to have that chance.

These teaching experiences were the reason I quit classroom teaching, and the inspiration for the Australian Data Science Education Institute.

The more I worked in this space, the more I realised that the engagement with tech skills was a bonus. A pretty good bonus, bringing more girls and greater diversity into the students choosing the year 11 Computational Science elective, and sending kids into tech courses and careers who tell me they would never have considered it otherwise, but this was still just a bonus to the real impact, which is that using real datasets and working with unsolved problems actually teaches kids to be critical thinkers. To be rational sceptics. To be, in fact, heretics. Heretics who are unafraid to ask difficult questions, to challenge established "wisdom", and to create new solutions.

It's really hard to teach kids critical thinking skills when your toolkit consists of questions that all have right answers, curricula full of facts and straightforward procedures, and textbooks that leave kids floating on an uneasy sea of factoids, memorisation, and perfectly neat examples tied up with a bow.

If you're using a dataset that hasn't been analysed before (and they are easy to find online – I could hit a dozen with a bread roll from where I'm sitting), then you can't know the answers in advance. This means you have to find ways of testing your results; not seeking to verify them, but seeking to prove them wrong if you possibly can. Suddenly the emphasis shifts from regurgitating facts and getting the right answer to exploring, questioning, and testing.

When I teach kids to explore a dataset, the first question we ask is "what's wrong with this data?", because no dataset is ever perfect. Yet when we use data in schools, to the limited extent we currently do, we use perfect datasets. We explicitly edit out any problems, and ensure that the answers we get are perfect and uncontestable, not confusing or unexpected.

The other night I was chatting to a dear friend who teaches Physics, Kathryn, while having a drink over video conferencing as befits the Covid19 social distancing rules (evidence based policy!). Kathryn was excitedly telling me about a thermodynamics experiment she conducted, recording her data with data loggers so that she could share it with her students and, if they couldn't conduct the experiment themselves during remote schooling, they could at least analyse the real data.

In the experiment, you put a cold object in contact with a hot one and track the temperature changes in each object. As expected, the hot one gets colder, and the cold one gets hotter, until they are both the same temperature and at equilibrium. Unexpectedly, though, Kathryn's data loggers recorded them crossing over after they hit what should have been equilibrium. So they reached the same temperature, but their temperature continued to change! The cold object continued to get hotter, and the hot object continued to get colder. According to the first law of thermodynamics, that shouldn't happen. I learned that law as a kid from a Flanders and Swann song, and it's pretty straightforward: *Heat cannot of itself pass from one body to a hotter body.*

So crossing over after they reach the same temperature should not happen. Kathryn was excited, because this was an opportunity to explore a weird result. Was it a problem with the calibration of the data loggers, or some unexpected interaction with the surrounding atmosphere? Or could it be that the loggers aren't that accurate, so the crossover wasn't real?

Unfortunately this was a year 11 class, and the year 11 Physics curriculum in Victoria contains a lot of content, so there wasn't time, or room in the course, to explore these unexpected results. For the kids, the best outcome would be to clean the data and give them a set of results that behaves exactly as expected.

Data that contradicts the "hard facts" they have learned will confuse the students, given that they have been taught that science is all hard facts and clear, predictable outcomes. And the priority for year 11 in almost any school is for students to maximise their results. Even more so in year 12. And results are a measure of students' recall of those hard facts, and application of nice, predictable formulae and concepts.

Students' ability to deal with unexpected results, to ask challenging questions, and to be sceptical, is not measured. If you don't measure it, you don't value it. Teachers get frustrated with students asking "will this be on the exam?" just when we get to something interesting, but we have explicitly and implicitly taught those very same students that numeric results are what matters.

Where to from here?

In this book, I want to show you how Data Science Education is key to nurturing a rationally sceptical, creative, ethical, problem solving population who can save the world.

I'm going to do that by looking at the problems we have in the Data Science and Technology communities today, and how those communities are shaping our world – problems and all – in Chapter 1: "Who's in Charge?"

Given that Data Science is in the driver's seat, taking us towards a future we are not yet equipped to understand, Chapter 2: "The Shape of the Future", talks about what the future *could* look like if everyone had enough data literacy to form evidence based policy, support high quality science, and have a say in the shape of our future.

Of course, if we want an evidence based society that treats science with respect, we need to understand how science actually works. Too often a change in our understanding of something – whether it's climate change, a virus, or our diet – leads us to think that science got it wrong. Scientists, however, know that this is how science progresses; by improving our understanding of complex systems. That means that sometimes what we think we know about science today turns out to be wrong tomorrow. This is science at its best. Unfortunately there is a perception in the wider community that science is solved. And science education reinforces that idea quite firmly. Chapter 3: "Science is Solved", looks at the way we (mis)understand science, and how we can fix it.

I'm then going to explore the issues with our current education system in more depth. There is no such thing as perfect data, yet we treat data with more reverence than it deserves. Our entire education system is built on the idea of being measurable, yet all too often "measurable" winds up being the opposite of "meaningful". Chapter 4: "Measurable or Meaningful, pick one," considers how we got here, and how we can create an education system that focuses on meaningful outcomes, and develops our students into rational, ethical heretics.

All of these goals require us to get comfortable with the idea of uncertainty. To be prepared to challenge the status quo, query accepted wisdom, and even to question our own findings. Chapter 5: "Accepting the Unexpected," focuses on why uncertainty is important, and how we can get comfortable with it, especially in education.

Why should you take my word for it? Chapter 6: "Projects with Impact," goes into detail about how Data Science projects work, with case studies from my own teaching, and Chapter 7 outlines templates for Data Science projects involving community projects and more global issues, with examples of units ADSEI has created right across the curriculum, from Humanities to STEM.

Finally, how do we get there from here? Chapter 8: "What now?" maps out what we need to do to overhaul our education system and raise all of our children to be rational heretics, so that they can understand the world, and then save it.

1

Who's in charge?

Class after class of my year eleven Computer Science students used to complain when we started the Usability unit of the course. "Why are you making us do this, Doc? We just want to code!" These kids wanted to learn straightforward technical skills, not mess around with complicated people stuff like usability, which is the study of how to make things easy to use, easy to understand, predictable, and reliable. Usability is everything modern technology is not, so it was understandable that Computer Science students didn't want anything to do with it.

By the end of the course they were still complaining, but now they were complaining about difficult door handles, obscurely unusable fan switches, and incomprehensible software. "You've broken me!" they'd cry. "I can't unsee this! Bad usability is EVERYWHERE! IT BUUUURRRRNNNNSSSSS!"

I take some pride in the fact that many of those students went on to study Software Engineering, Computer Science, Information Technology, and related degrees, and went out into the tech industry as people who see usability as crucially important. What worries me, though, is that many of them have come back to me years later, citing that unit as one of the most important things they ever did, and noting that they never encountered usability again in their studies.

Although I find it worrying, sadly it doesn't surprise me, because before I left academia I taught someone else's Introduction to Software Engineering course for a semester, and it didn't discuss usability at all. Indeed, the 650 page textbook for the course included less than half a page on the topic. So it comes as no surprise that it's not a major priority of tech courses, but it is deeply disturbing. Usability puts the users, not the technology, front and center of the design process. *What do users need? How do we help them?* Whereas the tech industry, for the most part, puts the technology first. *What can we build? How can we make money off it?*

Data Science suffers from the same problem. All too often, rather than saying *"What are the problems, and how can we use data to solve them?"* industry actually says *"What data do we have, and how can we monetise it?"*

Why is that? Who is charting the course and deciding what our future looks like? And how do we put society in charge of where technology is taking us?

Show me the data

Data, used well, can be a powerful force for good. Long before Data Science was even a term, Florence Nightingale was saving lives with data, by standardising the way sickness and mortality were recorded.

"Upon arriving at the British military hospital in Turkey in 1856, Nightingale was horrified at the hospital's conditions and a lack of clear hospital records.

Even the number of deaths was not recorded accurately. She soon discovered three different death registers existed, each giving a completely different account of the deaths among the soldiers. Using her statistical skills, Nightingale set to work to introduce new guidelines on how to record sickness and mortality across military hospitals.

This helped her better understand both the numbers and causes of deaths. Now, worldwide, there are similar standards for recording diseases, such as the International Classification of Diseases."[9] Excerpt from "The healing power of data: Florence Nightingale's true legacy" in The Conversation on May 14, 2020.

This simple step meant that Nightingale and her colleagues could actually start to understand and treat the conditions that were killing the soldiers. In particular, she quickly realised that the unsanitary conditions in hospitals were far more dangerous to the soldiers than the actual battlefields. Knowing this, vast improvements in treatment and survival rates immediately become possible.

Accurate, comprehensive data helps us truly understand a situation. It's essential for solving problems, and, crucially, for measuring how *well* we have solved them. The census, for example, gives us a clear picture of populations. It helps us to understand where we need to build infrastructure such as hospitals, schools, and public transport, because it gives us information about where people live, where they work, and what they do.

Unfortunately, data like this isn't always used in the formation of policy. Too often we see governments build infrastructure based on a different dataset – where it will buy them the most votes – rather than where it is most needed. For example, before the 2019 Federal election in Australia, the Morrison Government doled out grants to sporting organisations that were in marginal electorates[10], rather than following their own guidelines for how the grants were to be distributed. Some grants were made to organisations that had not even applied. Others were forced into the system by the minister concerned, Bridget McKenzie, after the committee that was supposed to make the decisions had created a list of successful applicants[11].

This scandal, which became known as Sports Rorts, had very little fallout for the government. It seems that we take it for granted that governments will operate in their own interests, rather than according to the evidence of what will build a better world.

Being able to understand and critically evaluate the relevant data is crucial if we want to hold governments to account for such decisions.

But it's not just governments who use data in ethically dubious ways to shape our worlds. How much do you know about how Facebook uses your data? Or Instagram (incidentally, owned by Facebook). Or Whatsapp (also owned by Facebook). Or Giphy (also owned by Facebook... wait, I'm seeing a trend). Does it really matter what data they collect about you, and how they use it? Do you have control over who sees your data?

Maybe you don't use Facebook because you don't trust them. Do you have an email account on a free internet service? Everything you put or receive in that email is available to the provider company for data mining. Do you use a rewards card with a store? How about a credit card? All of your purchases are recorded and data mined. Web browser? Your activity is being tracked. Mobile phone? Many of the apps are harvesting data about your behaviour. Have you installed the covidsafe app? How much do you know about what it records about you?

My friend Riley Taylor recently sent me an article, concerned about Facebook's purchase of Giphy[12]. At first I wasn't too concerned. What can any company really learn about me based on which gifs I insert into my conversations? To be honest, I don't use gifs much anyway. I'm all about words, not pictures. (Every image in this book was one I had to force myself to use!). But then he raised the possibility of tracking other activity using a gif.

"There may be no data/analytics tracking on a particular forum, but then someone posts a giphy gif and facebook injects their analytics code thereby giving them access to track everything a person is doing on that page."

Suddenly, because someone else used a gif, Facebook can see everything I do on that page? How much access does that give them? Does it give them my whole conversation if someone uses a gif in an email? I don't know. Do you? I have a PhD in Computer Science, but you could fill whole universes with quirks of the internet that I don't know about. How much does the average user know about what is done with their data, and where that might lead?

We're terrified, often, of new technology and the implications of it. Many years ago – 15 years or more – I owed my parents some money for tickets they had bought. I was planning to pay it back by direct depositing into their account, using my internet banking. Before I found time to do that, though, my elderly dad called me in a panic. They had just received a phishing email pretending to be from the small, local bank I use. They were worried this was because I had put their details on the internet. None of my reassurance sank in, and they demanded that I not use the internet, but pay them in cash next time we met.

I couldn't persuade them that the phishing email was purely a coincidence (I hadn't even tried to repay them at this point). I also pointed out that their banking details were already online, because their banks – and probably every organisation they have ever dealt with – were online. It didn't help. They were afraid of every internet transaction, so they opted out. 15 years ago you could just about get away with that. These days, especially during the pandemic, not being online isn't a realistic option anymore. Those who don't have internet access, or whose access is restricted to phones with very little data allowance, are severely disadvantaged.

I had a boss once who was deeply interested in Twitter, but he wouldn't make himself an account because he had read too many horror stories of how seemingly innocent tweets could destroy your life. He thought that anything he tweeted could ruin him. It never occurred to him that he could use Twitter without ever sending a tweet himself. He was afraid to go near the site at all. In his mind, using Twitter had become a sure path to ruin. He was afraid of the internet just like my parents, even though he was a different generation.

The trouble is that it's very difficult to know what we actually should be afraid of, and what is really low risk. We lack the knowledge, and the past experience, to realistically understand the cost-benefit analysis of engaging with these data hungry systems. This means we are at the mercy of companies trying to sell us things – or simply trying to sell *us*, in the form of all of the data they can collect about us.

Meanwhile, all kinds of disturbingly effective tactics are used to manipulate us into giving up our data. I first got a Facebook account in order to see a friend's pictures. I intended to delete it immediately afterwards, but then a long lost relative connected with me there, and I was trapped. Facebook finding friends to connect you with sounds positive, but it's not designed to make you feel more connected. It's designed to keep you on Facebook. Games, shopping, quizzes, celebrity pages – every interaction you have with Facebook gives them a little more information about you, and a little more data they can sell. They want to keep you on the system as much as possible.

When the Australian Government brought in MyHealthRecord, an electronic system designed to connect up patients' health information, making it easier for healthcare records to follow a patient from one service provider to another, around 10% of Australians opted out[13]. That's

90% who didn't, but how many of them had ever even heard of it, much less understood any of the issues around it? Of the 10% who did opt out, many had grave concerns not so much about how the health system would use the data, but about the government's ability to keep that data private.

When the Australian Government brought out the CovidSafe contact tracing app to try to manage the pandemic, around 5 million Australians downloaded it almost immediately, even though it quickly became apparent that it didn't really work.[14] In particular, it did not work on iPhones unless the phone was unlocked and the app was open and in the foreground – in other words, it didn't work if the user was on Tiktok, had their phone in their bag, was on a call, etc. In that case the message was clear – Prime Minister Scott Morrison explicitly said that downloading the app was the ticket to easing restrictions.[15]

Given that most Australians were under fairly strong restrictions around where they could go and what they could do at the time, that's pretty powerful coercion! Incidentally, it is now widely accepted that the app does not work. It has not been useful in contact tracing, and is frequently publicly disparaged. Victorian Health Minister Martin Foley was recently asked whether covidsafe had helped with contact tracing in the latest outbreak, and replied, *"No. Not to my knowledge, and I'm sure in such a rare event it would have been brought to my attention."*

Much of our response to technology and Data Science is emotional, and we are manipulated into responding the way governments and companies want us to, but how much do we actually know about the way any of it works, and how it can be used for or against us?

How Data Science & Technology are shaping our world

Data Science – the collection, analysis, and communication of data – is shaping our world and our future, yet in many ways it's going horribly off course. We tend to assume that Data Science is a precise science with fixed and reliable outcomes. Unfortunately, Data Science is incredibly vulnerable to human bias and fallibility. When data is collected, processed, analysed, and communicated, things can, and often do, go so wrong as to completely skew our understanding of the world.

We have a wholly unwarranted faith in computers, technology, and data, that leads to us placing our trust in systems that cannot possibly do what they promise. It's not the data itself that's the problem. It's the assumptions we make about it. We assume, for example, that it includes everyone, or applies the same way to every person in every situation, never changes, or has no other aspects that are not captured by the data we have. It's the way we use it, and our collective failure to question it, that cause problems.

Systems are cropping up in almost every field that promise to use a combination of data and machine learning, or "artificial intelligence", to solve intractable problems, such as figuring out who to hire, predicting where crimes are likely to happen[16], calculating what someone's exam result would have been if covid had not prevented them from sitting exams[17], detecting and diagnosing diseases accurately, or working out the appropriate credit limit for a person applying for a credit card. All too often these systems are treated as magic bullets, rather than as new, untested, unproven systems that may or may not actually solve the problems they claim to solve.

Consider the term "artificial intelligence" itself. Because clearly an intelligent system is a good thing, right? And an intelligent computer must be objective and unbiased and far more accurate than a human being, because computers are always right.

The problem is that no-one has ever actually developed an intelligent system. I used to spend hours with my year 11 Computer Science class debating the nature of intelligence. It's not easy to define, and you can debate around the edges endlessly, but one characteristic that is universally agreed on is adaptability: To be considered intelligent, an entity must be able to apply existing knowledge to new scenarios. To date no-one has developed a system that can do this.

What we can do, sometimes, is develop systems that are phenomenally good at very specific tasks. For example, Professor Regina Barzilay at MIT developed a system that was better than a radiologist at detecting and diagnosing breast cancer.[18]

However, that same system would completely ignore any other pathology in the breast. It can give you a clear answer and probability of cancer, but it can't tell you there's a cyst, a blocked duct, an infection, or anything else that might be causing issues. I read many, many articles about Professor Barzilay's work, all trumpeting the accuracy of its findings, and the improvement on radiologists' ability to diagnose cancer. Because that's what the testing of the system looked for.

Not a single article pointed out that if there was any other issue, the system would not be able to identify it, or even flag it as anomalous. In essence, the system assumed that no cancer meant no problem. Now, I have certainly encountered doctors who follow the same assumption – "You don't have the problem I specialise in, therefore you don't have a problem." But it's worrying to see that encoded in computer systems that might wind up never interacting with a human being who can say "hey, there is no cancer, it's true, but there is a serious problem here that still needs to be treated."

This is clearly not intelligence. It's a really important advance – improved diagnosis of breast cancer is obviously great news for patients. But it can't replace radiologists entirely, and it's not intelligent.

In 1950 Alan Turing speculated about a way of identifying an intelligent system. Now known as the Turing Test, the idea was that if you were having a text-based conversation with a system or a person in another room, and you couldn't tell, when talking to the system, whether you were talking to the human being or the computer, the system could be classified as intelligent.

But even this definition is far narrower than we really need in order to define intelligence. A chatbot might be able to hold a plausible on-screen conversation, but be completely unable to identify a dog as an animal, to recognise the sound of rain, or to know when it's a good idea to put a raincoat on. It might be able to hold a sensible conversation with one person at a time, but completely fall in a heap if another person joins in. Where a human being could infer a lot about a person from the conversation (sometimes incorrectly!), even an AI that can chat and appear intelligent is probably not constructing a mental model of the person it's talking to.

So we can't create intelligent systems, but we can create systems that are incredibly good at one thing, under highly constrained and very specific conditions. Another issue with these systems, though, is that they're not always doing what we think they're doing.

There is a well known story which is probably apocryphal, but that makes the point very effectively. Someone in an army somewhere was trying to develop an AI to detect tanks in an image. They hired a tank, took it to a field, and spent the morning taking photos with the tank in various parts of the image, and various settings. They then sent the tank back and spent the afternoon taking the same pictures without the tank.

To their great delight the AI they trained on these images was 100% successful in identifying which pictures contained a tank, or parts of a tank. Fortunately one team member was deeply suspicious of this – you never get 100% success. It seemed too good to be true, so they went back and looked at the training data. It turns out that the sun had been out in the morning, when the tank photos were taken, and it was cloudy in the afternoon. A little more testing and the sad truth was clear. Their excitingly 'intelligent' tank-detection system was successfully detecting whether or not the sun was out.

In another example, a pneumonia detection AI in America that showed great accuracy didn't actually learn to detect pneumonia in X-Rays. Instead it learned predict pneumonia rates based on conditions that were specific to the institution it was trained at:

"Researchers at the Icahn School of Medicine at Mount Sinai were deeply puzzled by a discrepancy in the performance of a deep-learning algorithm they had developed to identify pneumonia in lung x-rays. It performed with greater than 90 percent accuracy on x-rays produced at Mount Sinai but was far less accurate with scans from other institutions. They eventually figured out that instead of just analyzing the images, the algorithm was also factoring in the odds of a positive finding based on how common pneumonia was at each institution – not something they expected or wanted the program to do." [19]

This is an issue that is, to some extent, baked into machine learning algorithms. By their very nature, they are set loose on a dataset and expected to detect patterns. The dataset almost certainly has its own flaws in terms of the bias that stems from unrepresentative data or small sample sizes, missing data, and overrepresentation of particular groups, among other issues. And the algorithms are very good at detecting patterns. In those cases where they detect patterns we did not yet know about – for example, particular cell shapes that are precursors to cancer – they are phenomenally useful. But it's sometimes very difficult, without rigorous and extensive testing, to know whether the patterns it is detecting are the right ones – whether it is detecting a tank or the sunshine.

The field of AI as a whole is becoming more aware of these issues, and starting to shine a spotlight on them, in the hopes of improving the accuracy, reliability, and credibility of AI. Unfortunately in both academia and business there are rewards for speed and impressive results, and actual penalties for taking the time to do the kind of sceptical, comprehensive testing that might put your results in doubt and see your funding withdrawn, or your journal paper rejected. As a result, we see systems rushed to market and widely deployed that perhaps could have done with a little more rigorous testing (or, indeed, *any*).

Consider, for a moment, the example of HireVue AI[20]. They're a Human Resources Tech company that uses artificial intelligence to select or reject candidates in job interviews based on... well, nobody actually knows.

They say it's a machine, therefore it's without bias. And sceptics like us might scoff, but over 100 companies are already using it, including big companies like Hilton and Unilever.

According to Nathan Mondragon, HireVue's chief industrial-organizational psychologist, "Humans are inconsistent by nature. They inject their subjectivity into the evaluations, But AI can database what the human processes in an interview, without bias. ... And humans are now believing in machine decisions over human feedback."

Leaving aside, for a moment, the use of database as a verb (because it makes me flinch, which interferes with my typing no end), this is really all kinds of disturbing. They are making grandiose claims for their system, but can't, or won't, actually explain how it is making its decisions.

They say that the system employs "superhuman precision and impartiality to zero in on an ideal employee, picking up on telltale clues a recruiter might miss." But they won't, or can't, tell us what kinds of clues they might be.

Of course, HireVue won't tell us how their algorithm works – in part to protect trade secrets, and in part because they don't really know. We assume that an "artificially intelligent" piece of software can do things that we can't, but we don't know how it does it, or even whether it can actually be done.

HireVue's system is trained on the results of previous interviews – which, of course, means that any bias that existed in the human system are now encoded and potentially magnified in the artificially intelligent system. Far from being objective and unbiased, these systems are now making decisions that can't be queried, using data that is fundamentally biased, all wrapped in a cloak of mystique that makes it both more believable and less trustworthy than the original biased humans it is intended to make redundant.

Niels Wouters from the University of Melbourne expresses it succinctly when he points out that this kind of magical thinking risks codifying existing biases.

"Even though Computer Scientists may show off how accurate and reliable their machinery is, the concern is that we are only automating old fashioned and concerning human classification techniques. Think eugenics, social darwinism, and phrenology."

Of course, artificially intelligent algorithms are not the first examples of flawed Data Science. When Functional Magnetic Resonance Imaging (fMRI) scans first became possible in 1990, the field of neuroscience was wildly excited. Here was a scan that could highlight active areas of the brain in a live subject. For the first time it was possible to show brain activity while a person was actually doing a task. An fMRI image, though, is not a photograph like an XRay. It's a visualisation of a whole lot of statistical wrangling of the data recorded inside the MRI machine. Gina Rippon describes it beautifully in The Gendered Brain[21]:

"The production or 'construction' of a brain image, either of a single individual or a group, requires a multi-layered hierarchy of decisions, about how to 'clean' the raw data, how to smooth individual anatomical differences, how to 'warp' brain characteristics to fit a template brain. The allocation of colours to different types of changes identified is actually a statistical procedure. So the flickering colours that move across the grey and white tundra of the brain as someone views a coke commercial are not equivalent to a time-lapsed sunset but reflect some thresholding decisions made by a brain imager." Gina Rippon, The Gendered Brain.

The credibility of fMRI scans took a serious hit in 2009 when sceptics Craig M. Bennett, Abigail A. Baird, Michael B. Miller, and George L. Wolford decided to prove once and for all that the hype over fMRI was dubious at best, dangerous at worst [22]. They put a dead salmon in an fMRI machine to see if it would give results. It's worth reading the poster describing the study for the entertainment value of dry, scientific descriptions of patently ludicrous events:

Subject. One mature Atlantic Salmon (Salmo salar) participated in the fMRI study. The salmon was approximately 18 inches long, weighed 3.8 lbs, and was not alive at the time of scanning.

Task. The task administered to the salmon involved completing an open-ended mentalizing task. The salmon was shown a series of photographs depicting human individuals in social situations with a specified emotional valence. The salmon was asked to determine what emotion the individual in the photo must have been experiencing

As the team feared, the dead salmon's fMRI results showed "brain activity" when standard processing techniques were used, thus demonstrating that they either had a zombie salmon on their hands, or that the usual techniques had some issues that probably needed correcting.

Even relatively simple spreadsheets can have significant issues causing widespread problems. In 2010, Harvard Economists Carmen Reinhart and Kenneth Rogoff famously published a paper showing that economic growth slows by an average of 0.1% if a country's debt reaches 90% of its gross domestic product (GDP). This finding was used by politicians and financial institutions to justify sweeping austerity measures.[23]

It wasn't until 2013 that a graduate student by the name of Thomas Herndon obtained the original spreadsheet and found that Reinhart and Rogoff had not included all of the countries they thought they had in their analysis, because they had failed to highlight the entire row, thus including just 15 of the 20 reported countries.

When all 20 countries were included, the 0.1% decrease became a 2.2% increase. There are so many issues with this study that it's hard to know where to begin. It's a nice, neat illustration of the need for double-checking your calculations, but it's also a powerful cautionary tale. Why would global economic policy be based on a study of just 20 economies anyway? If you added another 5 countries, it might well produce yet another, radically different result.

One of the problems here is that humans are deeply susceptible to a thing called confirmation bias, which means that we look for evidence to support our existing beliefs, and tend to rule out or ignore any evidence that contradicts them. This is one of the reasons that climate scepticism is a thing – if you want to believe that the climate is fine, and there is no need for change, you can easily find people saying things that agree with you, ignore anything contradictory, and stay safely cocooned in your ignorance.

In the case of Reinhart and Rogoff, it may well be that they stopped as soon as they found the result they were looking for. Which is not really surprising since, as we'll see in chapter 3, that is the way we are taught to do 'research' from a very young age. "Do this experiment. Find this known result." This is one of the crucial reasons why we need to teach students right from the start to question and challenge their own results.

So these are errors, made by people with debatably good intentions. What about wildly unethical uses of Data Science? In May 2020 the state of Georgia in the United States of America published this graph on its website to show that its cases of covid19 were dropping.[24]

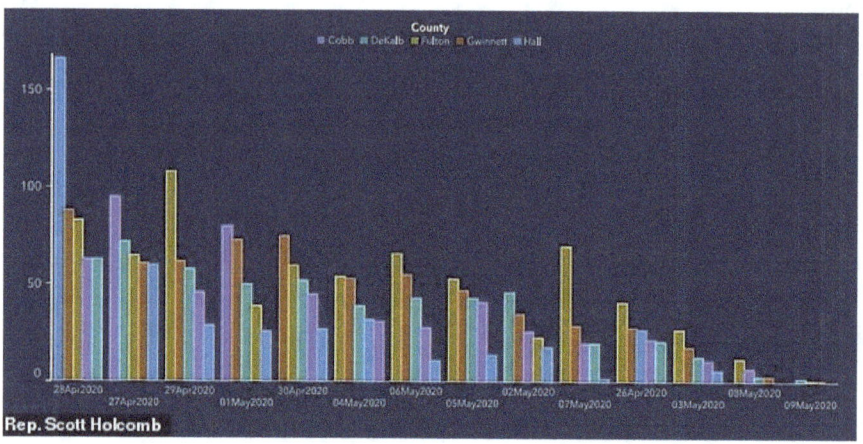

Georgia's graph of the counties with the greatest number of confirmed Covid19 cases over time

It does, indeed, look like a graph that shows decreasing numbers. Unfortunately, close examination reveals that the dates along the bottom of the graph are not in order. In addition, though you can't tell this from the graph without having the raw data as well, the counties were not all in the same order. In other words, the little cluster of columns labelled the first of May, which came after the 29th of April but before the 30th in the strangest bit of time travel I've seen in a while (and I watch Doctor Who), was not necessarily all data for the first of May. Some of the counties were sorted differently, so that it's entirely possible that only one of the columns clustered around 01 May 2020 was actually from May the first.

It's difficult to see how this could have been accidental. It's a wildly misleading and suspect way to present the data, clearly intended to present a particular picture of events that bears no resemblance to reality. Below is a more conventional presentation of the actual data. You will note it is not nearly as neat as the, um, *altered* graph. That's often a very strong clue that data is not real, or the representation is invalid. Because neat data very rarely happens in real life. Real data is messy, complicated, and difficult.

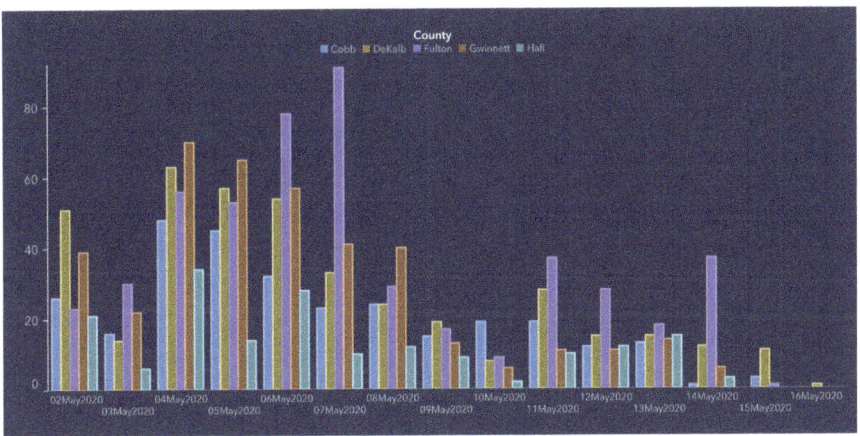

Accurate graph with the cases in chronological order

For a really disturbing misuse of data, look no further than employee monitoring software, which has become increasingly popular in Australia during the unprecedented level of working from home triggered by covid19.[25]

You might think it's not that different from checking your employees' screens as you walk past in an open plan office, until you consider the introduction of the insidious "productivity score".[26]

It's a potentially attractive idea – who wouldn't want to know how productive their employees were being? Unfortunately any meaningful measure of productivity has to be, at least in part, a qualitative judgement. To measure productivity you must also evaluate quality of work rather than just quantity, and that is nearly always impossible to assess as a simple metric, particularly by monitoring a person's computer-based activity. Churning out a vast amount of work rapidly will be assessed by a quantitative measure as extremely productive, regardless of whether the work is actually of usable quality.

A quantitative measure of productivity cannot capture the one earth-shattering idea that revolutionises the project, or solves a thorny problem that has been delaying progress for months. It may, in fact, penalise thinking time, as it does not show up as active computer use (or may even show up as idly scrolling social media in some cases – which, incidentally, is where the inspiration for this section came from). Interestingly, recent research at Stanford shows that quantifying productivity for complex work actually demotivates workers and reduces productivity.[27]

Microsoft's version of the productivity score claims to work by, among other things, "quantifying how people collaborate on content". Of course, they do this automagically – and I say automagically rather than automatically because of the magical thinking that underpins the idea that this is even a thing you can measure. There are whole fields of research on how to measure collaboration, and much disagreement on whether it is even possible. There is no evidence that automated measurement of collaboration comes up with anything at all meaningful. And, of course, it is not transparent, because no-one will tell you how the algorithm works to calculate the score.

Imagine two people collaborating on a document. If one works on section A and the other on section B, is that, in fact, collaboration? How would it score? If they have a terribly polite and friendly conversation in the chat section of the editor that completely fails to address the difficult section of the document that no-one knows how to write, would an automated collaboration analyser define that as good collaboration, because the tone of the conversation was friendly? What if one of the employees was rather sweary, but in a good natured way, and the other mock-scolding her for it. How would the algorithm interpret that?

Now take these "measures" of collaboration and productivity and assume they get used for performance reviews, or to figure out which employees to get rid of in the next round of cuts, or for promotion purposes. How confident would you be that the best employee is being identified by this system?

When I choose a doctor, I tend to appreciate the ones who often run late, because they take the time to talk to their patients and really get to the bottom of any issues, rather than churning them through the door at the highest rate. Yet a productivity algorithm would probably rank those doctors as unproductive, even if they are actually more successful at treating their patients (which is much more complex to measure).

The alarming problem is that people believe in numbers, and they believe them to be reliable, objective, and true. This conversation with my friend Gretchen Scott about the Microsoft Productivity Score sums it up magnificently:

Gretchen: But people will LOVE it and believe it because numbers are true and real.

Me: Yes. And objective.

Gretchen: Totally objective, unemotive and rational. No bias or hysteria there. It's almost like we view numbers as men.

I love this analogy, because it makes as little sense to say that data is objective, rational, and unbiased, as it does to say that men are objective, rational, and unbiased. Human beings are fundamentally biased, irrational, and subjective creatures, whether male or female. Yet because people have this blind faith in the truth and objectivity of numbers, measures of fundamentally unmeasurable things are increasingly shaping our society.

This all highlights an issue I've seen raised again and again in works like *Weapons of Math Destruction*, *Made by Humans*, and *Automating Inequality*[28] – that people believe in numbers, computers, and algorithms over other people, even when they've been explicitly told those systems are broken. And I have a story about that, too.

Niels Wouters, a researcher at Melbourne University, some years ago designed a system called Biometric Mirror[29]. It was a deliberately simple, utterly naive machine learning system that took a picture of the user's face and then claimed to be able to tell a whole lot about the person, just from that picture. The system spat out a rating of ethnicity, gender, sexual preference, attractiveness, trustworthiness, etc. Niels created the system to start a conversation with people about how transparently ludicrous it was to believe a system that does this. So he set up booths where people would come, have a photo taken, and read all of this obviously false information about themselves, and then have a conversation about trust and ethics and the issues with Artificial Intelligence. So far so good. A noble goal. But there are two postscripts to this story that are horrifying in their implications.

First of all, Niels would overhear people walking away from the display, having had the conversation about how obviously false the "conclusions" drawn by the system were, saying rather sadly: *"But it's a computer, it must be right, and it doesn't think I'm attractive..."*

And secondly, after speaking publicly about all of the issues with Biometric mirror, Niels was contacted by HR companies wanting to buy it...

We'll talk more about measuring the unmeasurable in chapter 4.

For now, the lesson we can take from all of these examples is that we are increasingly ceding control over important parts of our lives – from employee hiring processes to healthcare – to algorithms and calculations that are often not transparent, not properly tested, and that don't necessarily take all of the people potentially impacted by them into account. They are, in short, thoroughly undeserving of the faith we have in them.

And that leads us to an important question: If Technology and Data Science are controlling us, who's controlling them?

REPRESENTATION & DIVERSITY IN TECHNOLOGY & DATA SCIENCE

I have a dear friend, Sarah, who knows I love chocolate, so when she comes over she always brings some for me. Unfortunately, it's nearly always something I can't eat, because it contains gluten. There are plenty of types of chocolate I *can* eat, but Sarah, I suspect, brings her favourites, wanting to give me something she knows is great. It's just unlucky that I have coeliac disease, so can't eat any gluten at all. Sarah doesn't have coeliac disease, so the idea of avoiding gluten just doesn't cross her mind. She's not used to thinking about it.

A lot of restaurants say they provide gluten free food. Unfortunately most of them don't. It might be ok for non-coeliacs who just want to "avoid" gluten, but for coeliacs who are highly sensitive to cross contamination, if "gluten free" food is cooked on the same surface, or in the same oil, or handled with the same utensils as food containing gluten, it's no longer gluten free, so it's not safe for us. Invariably when I find a restaurant that's really good at doing proper gluten free food and has a wide range of options for me, it's because they, or someone they're close to, has coeliac disease, so gluten is front and centre of their thinking whenever they cook.

Why am I talking about gluten in a book about Data Science education? Because the tech industry suffers from the same kind of ignorance. Not about gluten, but about anyone who is different in some way from the people who developed the system. It's really easy to develop a system that works well for yourself. It's much, *much* harder to develop a system that works for anyone, and *everyone*, who is different to you.

Perhaps they are physically different – say, they have larger fingers or less manual dexterity (due to arthritis, maybe), they are blind, colour blind, or just have trouble reading small print – or they are mentally different – say, unfamiliar with the technology, dyslexic, speak a different language, don't know the same terminology as you, or simply do things differently.

If it's a form of difference you're not super familiar with, and used to incorporating into your life and work, you're very likely to forget to take it into account when you design new technology. This is how we wind up with, among other things, iPad based point of sale systems that are unusable for blind people – because the developers almost certainly did not have any blind friends or family. It was a use case that they probably never considered at all.

iPad based payment systems mean you only have to tap your card, and then you type in your PIN on the touch screen of the tablet. But there is simply no way for a blind person to find the "buttons" that represent the numbers in the PIN. They are not tactile. They don't actually exist. They are purely areas on the screen that are visually delineated. For a blind person the standard Point of Sale device which has raised buttons allows them to feel the keypad and figure out where the numbers are, so that they can type their PIN. An iPad Point of Sale device can do a fancy graphic of the receipt printing out, look pretty, and be lovely and easy to operate for most people, but it is simply not possible for a blind person to use if they need to type their PIN.[30]

This is one of the key reasons why diversity – of all types, not merely gender – is crucial in technology teams, whether software, hardware, or Data Science. Because when we are developing systems for everyone, we need to be sure everyone can use them, or be represented by them. Which means our teams need to reflect society as a whole.

At the moment, technology teams don't even come close to reflecting the gender balance of our world, and gender is the one thing that's often measured. If we're talking cultural background, age, physical ability, native language, or any other form of difference... well... the truth is that for the most part we are *not* talking about it. We're certainly not *measuring* it. And what we measure (or don't) speaks volumes about what we value.

So here we are, in a world which is being increasingly shaped by Technology and Data Science. Startup companies are tiny fledgling businesses one moment, and the next a small percentage of them will be highly influential, disruptive organisations the size of Uber, Facebook, Google, or Atlassian. Chances are that if you're not in the technology field you may not have heard of Atlassian – an Australian software company that makes software development and collaboration tools – but it has a classic startup story. Two guys meet at university, start a company with the sole aim of not working in a mundane job, and wind up billionaires. It's the tech industry dream. Atlassian was the first Australian company to become a "unicorn" – a privately held startup valued at over one billion US dollars. In July 2019 it was valued at over 50 billion.

Not every startup will be successful, of course. Most won't. But startups are the creatures that are increasingly disrupting and shaping our world. If you look at the statistics published by four of the biggest: Facebook, Google, Atlassian, and Uber, you find distressingly similar trends: Women in Senior Leadership roles: from a high of 30% at Facebook and Atlassian to a low of 26.7% at Google. Women in technical roles: a shatteringly low *maximum* of 23.6% at Google down to a minimum of 20% at Atlassian. Women overall: 40.9% at Uber down to 30% at Atlassian. Gender, of course, is the low hanging fruit of diversity issues. It's easy to see, easy to measure, easy to focus on. But, apparently, still difficult to fix. All four companies spend a lot of time talking about how important workforce diversity is to them, but their numbers are dreadful.

These companies have, in tech terms, been around for a while. Surely young companies will be better at this, right?

Sadly, it's not looking so great. I analysed a 2018 article in The Martec titled "100 Australian Startups to Watch in 2018" and counted male presenting and female presenting founders. (None were identified in the article as gender diverse.) I found that, of the 121 companies listed (apparently the authors need a little Data Science education themselves), only 10.5% of founders were women. That leaves 89.5% men. 98 of the 121 companies – 81% – had only male founders. These startups, which are busily shaping our future, do not represent us. Sure, that's founders, not employees. But it's significant. The people leading the charge to the future, determining the way our lives will look, are mostly men. The photos were disturbingly interchangeable.

And it's this homogeneity and lack of diverse perspectives that leads to period tracker apps that assume the length of everyone's period fits within a standard range of days. (In case you are in any doubt, there is almost infinite variation, sometimes for the same person.) One friend had her period tracker tell her that her period was invalid. It makes for a good story to be told your period is invalid, but imagine being told your child's medical history or schooling was invalid? Or their gender. Or their name. I've been told my identity is invalid because of systems that can't connect M‘Iver with MC IVER. Not that MC IVER is my name, but it's the only way some databases can deal with the little c, capital I situation. (and that's not counting the databases that think I am McLever, MacLver, or MacGyver).

In one example even I didn't think of, the contact tracing apps for covid19 assume you have your phone on you at all times, because they try to estimate how far away you are from other people. If your phone is in your bag some distance away, accurate contact tracing becomes impossible, because it's measuring the distance to your bag, which probably can't catch or spread covid. Which pretty much assumes you have pockets. Who doesn't have pockets?? Oh... women.

Or think of the microphones used for public speaking events. At one talk I gave recently I had to reconfigure my clothing – turning my scarf into a belt – just to allow the microphone to be attached. Because it was designed to be used with a man's suit, with a battery pack that goes in a pocket or clips to a belt, and a microphone that clips to the shirt between the buttons. If you are wearing a dress without a belt and without a conveniently clippable neckline, these kinds of technology are a pointed reminder that you don't belong. I mean, who could have predicted a speaker wearing a *dress*? (Even if you happen to be wearing a shirt, women's shirts button on the other side to men's, which means the microphone, usually fixed to the clip, will be pointing downwards on a female shirt, instead of upwards to capture the voice properly.)

Lack of diversity leads to explicitly exclusionary things such as drop down menus that allow you to put male or female, but not non-binary or gender diverse. It also means fewer people prepared to ask "hey, why are we collecting gender information on a page that sells cake *anyway*?"

If we want technology to take us all forward into a future that works for everyone, then the tech industry has to reflect and represent all of us. Male, female, gender diverse, blind, deaf, disabled, tall, short. All races & cultural backgrounds, all income levels, all states of housing – from mansion to no house at all.

Because technology is bound to be designed to work ideally for the designers. So at least *some* of the designers had better be like us!

When you narrow the focus from the technology industry as a whole down to Data Science, you find systems that are increasingly being used to determine health care, education, even the functioning of our democracies. If those systems are consistently being designed by and for white men between the ages of 20 and 45, we may find it's not just our periods that are suddenly deemed invalid.

DEMOCRACY OR OLIGARCHY – WHO IS MAKING THE DECISIONS, AND WHO PUT THEM IN CHARGE?

When we vote in an election, whether we are thrilled with the outcome or not, at least we know the people have spoken. Allegations of interfering with elections or rigging votes are serious and disturbing precisely because a free vote is so important to the functioning of a democracy. But increasingly our society is not planned and shaped by governments. It's planned and shaped by disruptive tech startups that can circumvent labor, privacy, and environmental laws by using new technologies in ways that legislation can't keep pace with.

Uber, together with all of its dubious progeny and imitators, has changed a significant proportion of our workforce into an insecure, underpaid, exploited gig economy, and governments have failed to stop it. A recent study in the US found that over one third of workers participated in the gig economy, and that number is constantly growing. [31]

Workers in the gig economy have no sick leave, no holidays, and often don't even make minimum wages. In countries like the USA where health insurance is mostly tied to work, gig economy workers don't get that, either. They are contractors, not employees, so they are not covered by employment law in any meaningful way, leaving them terribly exposed to exploitation and poverty. This is how Uber can offer you cheaper rides than taxis – because it does not pay its workers much at all.

Facebook has enabled election interference in Brexit, the USA presidential election, and the Australian election[32]. And though it storms and blusters, Facebook's goal is not to create a better society. Facebook's goal is to make money for Facebook, and it does that by increasing engagement with its users. Because election interference comes in the form of trolls that are highly sophisticated at creating engagement, it's actually pretty good for Facebook[33]. The same goes for Twitter.

These companies are having a significant impact on our society, yet they have not been elected, and they cannot be turfed out of office if we don't like what they are doing to us. "Elected" to these positions of power by a combination of technical know-how, business acumen, and a significant helping of luck, people like Mark Zuckerberg of Facebook, Jack Dorsey of Twitter, and Mike Cannon-Brookes and Scott Farquhar of Atlassian, have become de facto rulers of our future.

We are not even in a position to question them, because too few of us have any idea how the technology works, or what Data Science can – and cannot – be reasonably expected to do. Our legislators, largely comprised of lawyers, businessmen (and I use the term "BusinessMen" deliberately, as we really haven't achieved political diversity yet, either!), [34]unionists, and former political staffers, have not got enough Data Literacy to be able to pass appropriate, timely, and effective legislation. It is industry that is determining the direction we travel in, not government.

Our big problem is that, as a society, we understand so little about what Data Science is, how it works, and most importantly how it doesn't work, that we are in no position to even ask informed questions about it, much less to make evidence-based decisions about how we want it to shape our future. We desperately need to build a data literate, Data Science aware population that can take control of our future, and design an ethical, egalitarian, and healthy world that works for everyone.

Data Science Education can help show diverse students that they are capable of learning technology and Data Science skills, and, more importantly, that there are good reasons why they will want to. In the second half of the book we'll talk about the evidence for this, but for now I want to look at how much of our world is evidence based, and consider how society might look if we based our policies, decisions, and processes on evidence and data.

2

The shape of the future

Imagine, for a moment, a society driven by evidence. By data. A society that uses science, reason, and compassion to figure out the way forward. How would that be different from the world we see today?

Scientists and rational thinkers dream of a future where policy decisions are evidence based. From medicine to climate science, from welfare to pandemics, there is a clear difference between policy based on ideology and policy based on evidence. Too many decisions are made – in government, business, health, education, and elsewhere – on the basis of "logic", which is actually code for "this feels right to me", rather than "We've thought this through and looked at the evidence."

What we often call logic is basically intuition, which can be increasingly misleading for complex problems. It also gives bias an excellent entry point. Human beings are not nearly as rational as we'd like to think we are, but we are exceptionally good at rationalising our decisions[35]. We can make almost anything sound good. To spot our own biases, and those of others, we actually need to be trained to look for evidence, and to critically analyse the evidence we have.

We've seen the impact of evidence based action (and more speculative approaches) on the management of the spread of covid19. Those countries who heeded epidemiologists and responded quickly, such as New Zealand, South Korea, and Australia[36], have controlled the infection rate and protected their populations. Those who haven't listened to their experts, such as the UK and the USA, have seen spiking infection rates, mass deaths, and uncontrollable spread of the disease. As I write this (on November 12, 2020), the latest covid case numbers are 143,408 new cases detected in the USA, while Australia detected 8. It has been a striking example of how important data and expertise are in managing crises.

Yet we face an even more serious crisis in the shape of climate change[37], and here experts have been ignored, ridiculed, and even slandered in an attempt to postpone a response to what may be the greatest existential threat we have ever faced. There are, of course, differences between the spread of a virus with a rapidly rising and very visible death toll, and the slower, less tangible changes we have been making to our climate. This means that it is more important than ever that we have a population that is both scientifically and data literate. Who can see through the influence of fossil fuel lobbyists and the like, and demand rational, ethical, evidence based approaches to the escalating crises we face.

Clearly there is a need for evidence based decision making. Let's look at four topics – Medicine, Climate Science, Welfare, and Education – and examine the role of evidence in each area.

Evidence Based Medicine

When I was a kid, doctors were treated as demigods. Patients did what they were told, and trusted that medical treatment was always based on science and evidence. Despite a range of negative experiences with the medical profession over the last few years, my default response to health professionals is still one of trust, so I am always horrified when I look into the evidence base for particular treatments, or specific drugs, and discover the alarming lack of scientific rigour that underpins a lot of common medical treatments.

It is disturbing in itself that the term "Evidence Based Medicine" was first coined as late as 1991, by an academic by the name of Gordon Guyatt at McMaster University. It was not, initially, a way of practicing medicine. Instead, it was the name of a course designed to encourage medical students to make their practice more scientific.

If evidence based medicine was only just being talked about in the nineties, you have to wonder how medicine was practiced before that. Sadly, a startling amount of medical practice has historically been based on assumptions, untested theories, and arrogance. And much of it still is.

Consider the treatment of hip pain. In 2018, my daughter, Zoe, was diagnosed with acetabular retroversion and dysplasia, meaning her hip sockets were both too shallow, and facing the wrong way. She was sent to a physiotherapist for rehabilitation, to see whether her hip function could be fixed with the right strengthening exercises. We were exceptionally lucky that the physio she was sent to was Josh Heerey, who, at the time, was working on a PhD in hip problems. Being used to physios poking, manipulating, and making grand pronouncements on the basis of "feel", I was fascinated in that first appointment to see Josh using a dynamometer.

A dynamometer is a device for measuring force, and Josh used it to measure Zoe's strength in all directions. This meant that not only did Josh know for sure which muscles were weak and needed work, he was able to use the dynamometer on subsequent visits to measure Zoe's progress. Unfortunately Zoe's retroversion was severe enough that she needed surgery, but after 6 months of physio work, she was very strong, which made her recovery much easier. It also meant that her post-surgery rehabilitation was both scientific and effective, as Josh continued to measure her strength and prescribe exercises that directly targeted areas of weakness. After major hip surgery, Zoe is now running and jumping, with no sign of ever having had an issue, except for some trophy scars.

Meanwhile I started having hip pain, and went to a local physio. (Josh was quite some distance away, so I thought it would be quicker to see someone close by.) The local physio diagnosed bursitis, used a tens machine, ultrasound, heat treatment and massage, and after weeks of sessions I got precisely no improvement. In fact, if anything, I was getting worse. I asked Zoe's surgeon whether her condition was hereditary, and he ruefully confirmed that yes, it was likely her malformed hips were a genetic gift from her mother.

The bad news was that I was too old for the surgery that had helped Zoe. Before too long I was seeing Josh, and competing with Zoe to be the most obedient patient and do all of the exercises as prescribed. It was hard work, but within 6 months the surrounding muscles were strong, and I had no more hip pain. (Unfortunately I then had an insane 3 months of travel that trashed the *other* hip, so the dynamometer and I are currently close friends again.)

Traditionally the need for various radical and invasive hip surgeries has been determined from damage seen in X-Rays and MRIs. This was not based on studies showing a relationship between scans and pain or functional impairment, or indeed on studies of the effectiveness of the surgeries. It was simply a "logical deduction". A recent study of the relationship between hip pain and imaging results by Josh and his colleagues at LaTrobe University[38] found that there was actually no correlation between pathology seen in imaging and actual impairment of the hip. Imaging of people with no pain and no impairment showed similar levels of damage to imaging of people with pain and impairment, and there was no correlation between imaging results and hip function. It turns out that a lot of hip pain, my own included, can be effectively managed (and indeed banished) using physiotherapy.

And that's not an isolated finding. Various studies have found that different knee and back surgeries are no more effective than placebo[39] – in other words, if patients think they have had repair surgeries, but in fact have only had the incision and a bunch of experiences to make them *think* they've had surgery, their recovery is just as good as those who have had the actual surgery. Nonetheless, these surgeries continue to be performed, and described to patients as successful cures. Josh's evidence based approach to treating hip pain is not, unfortunately, the norm.

Vaginal mesh surgery has been at the centre of a worldwide scandal due to unexpected side effects experienced by as many as 14% of women, including disabling levels of pain, incontinence, and bleeding.[40] The mesh was introduced as a way to repair vaginal prolapse, but it quickly took off as a widespread treatment without any evidence of efficacy. Certainly it is difficult to determine the efficacy of a surgical procedure without performing it, but one of the significant concerns of the women involved is that they were given quite unjustified impressions of the risks and benefits of the surgery. In other words, they were told it would work, and not told that it was experimental and there were risks. There are standard procedures for performing trials and monitoring outcomes, with informed patient consent, that were often not followed.

Even stents placed in arteries to treat coronary heart disease are used without any evidence at all of efficacy. Once again, placebo procedures are just as effective. Eric Patashnik from Harvard says that as many as 50% of medical procedures in the United States may not be supported by evidence, and it's likely that the story in Australia is not greatly different.[41] Of course, that's just an estimate, because controlled trials and objective, scientific studies of efficacy and safety simply haven't been done for many of the treatments that we take for granted.

The story for medications is, if anything, more horrifying still. Until recently drug trials never included women[42], because our hormonal cycles were assumed to be too messy and complicated to yield nice, clear results. Anyway, the assumption was that we were just slightly smaller, differently shaped men. So drug trials were exclusively conducted on men, and typically young caucasian men at that, notwithstanding that we have known for decades that ethnicity can also have a significant impact on drug responses.

In fact, in the USA, where a significant proportion of drug trials are conducted, it was only in 1986 that it was "recommended" that women be included in drug trials, and not until 1993 that it was legislated that they must do so.

I was appalled when I read this, but relieved that things have changed. Unfortunately they have not changed as much as you might think. While women are, indeed, included in most drug trials now, the data is not disaggregated by sex – in other words, they don't look at male and female outcomes separately, even though many drugs behave quite differently in men and women. Aspirin, for example, improves outcomes in men with heart disease, but actually worsens them in women[43].

In 2019 I thought it would make a good High School Data Science project to compare studies that disaggregate results by sex with those that don't. When I examined over 100 recent studies on commonplace medications such as ibuprofen, panadol, antibiotics, and the like, though all included women in the study cohort, not a single study separated and analysed results by sex.

In addition, drug studies are not comprehensive. The goal of a drug company studying its own drug (and these account for the majority of drug trials) is to get the drug to market and make money from it. They use very select populations with no confounding issues or other health conditions, and they typically do the bare minimum of investigation. There is little incentive to discover whether the drug is less effective than other medications already on the market, or even to be sure it works at all. As Patashnik puts it, "Information about the comparative effectiveness of treatments is a public good, and the market won't provide the socially optimal level of this information without public subsidy." Market forces, in the absence of adequate regulation, tend to push for profit over social good, and the pharmaceutical industry is a clear example of this.

Even in cases where the data exists and is clear, such as in the use of surgical checklists to improve patient outcomes[44], doctors sometimes resist the findings. Checklists in surgery ensure that basic protocols such as handwashing, sterilisation, keeping track of sponges etc are followed. The evidence is strong that they can reduce complications, post-operative infections, and overall post-surgical mortality dramatically. Yet some surgeons consider them an unnecessary slur on their professionalism, despite the accumulated evidence of their effectiveness. The drive to have checklists adopted in all hospitals has used a combination of evidence gathering, story telling, and publicity, because evidence itself is not sufficient to change behaviour[45]. That theme will recur throughout this book.

Whether surgery, drugs, or other interventions are considered, it's clear that the current state of clinical medicine is not nearly as evidence based as we like to think.

Obviously the medical profession, in which we place our trust, and which we assume to be one of the most evidence-based, scientific disciplines, is subject to human bias, fallibility, and lack of logic, like all human endeavours. This means that billions of dollars worldwide are spent on useless, if not actively harmful, treatments. Not to mention the extraordinary patient trauma that can result from untreated, or mistreated conditions.

One of the fundamental issues with evidence based medicine is that it's impossible to know whether something will work in humans until it has been tried in a wide range of humans under a wide range of conditions. Sometimes even when trials have been comprehensive, the real world is sufficiently complex and varied that unexpected things happen. For example, it turns out that grapefruit juice interacts with some medications, including commonly used drugs like statins, antihistamines, contraceptive pills, and corticosteroids[46]. While it is common practice to ask patients what other drugs they are taking, in order to check for potentially harmful interactions, nobody would ever have thought to check on grapefruit juice as a potential risk factor.

We obviously can't make evidence based decisions in the absence of evidence. This means that when a new treatment is introduced, it is essential to monitor patients for an extended period, in order to identify common issues. This, in itself, is not straightforward, because a patient on a particular drug or who has had a particular surgery may have an adverse health event which has nothing to do with the drug or the survey. Fortunately, this is where Data Science has an effective role to play – if all patients are monitored and all health issues recorded for, say, two years after a drug treatment has been introduced, and everything about those patients and their health gets recorded, then Data Science can be used to detect patterns and put a stop to dangerous situations before the issues become widespread.

Unfortunately, our health & pharmaceutical industries are not setup to value this kind of process. Why would they pay for what must be an expensive set of trials and data collections that might wind up proving that a drug is dangerous and should be taken off the market? This would be bad for the immediate profit margins of the company that has just spent a fortune developing that drug and making it profitable.

In addition, there are complex issues around privacy and control of patient information that are not straightforward to solve. When My Health Record was introduced in Australia to allow patients to access their health information more easily regardless of which health provider they are seeing, over 2.5 million Australians opted out of the service, mostly due to privacy concerns.

Ultimately the issue is one of informed consent, both for data and for treatment. In the case of the vaginal mesh, patients were not informed that it was a new, unverified procedure with unknown risks[47]. Who knows how many would have gone ahead with the treatment if they had known? Plus adverse reactions were not recorded or reported, so the use of the mesh continued long after it should have been stopped. In the case of data, patients must be clearly informed when treatments are in a trial phase, and must be able to opt out of having their data collected. In addition, stringent controls have to be put into place to protect and anonymise (as far as that's actually possible) the data that does get collected.

Of course, informed consent also relies on patients being educated enough to understand the information that they are presented with, and on physicians being educated enough to communicate effectively!

In the case of vaginal mesh, many women felt abandoned by their doctors, and their concerns were ignored. Women who reported disabling pain, among other symptoms, were assumed to be overreacting or hysterical. Had a proper trial with effective monitoring been in place, the damage might have been fixable for many people who have, instead, been left with permanent disabilities.

Ultimately, when a patient is told "this is the most effective treatment we know of for your condition", they should be able to have faith that there is *evidence* for efficacy, and they should be able to find out what that evidence is very easily.

Evidence based medicine is clearly not straightforward, but there's no doubt we can do better.

Evidence Based Climate Response

We are used to talking about the Greenhouse Effect as a disaster in the making, which of course it is, but it's useful to know that, without the natural effect of Greenhouse gases, the earth would actually be uninhabitable. If we can understand the impact of carbon dioxide (CO2) concentrations, the implications of the man-made change in atmospheric CO2 become obvious and, to be honest, quite terrifying. In this section you will learn about the earth's natural temperature regulation, and how the atmospheric concentration of CO2 and other greenhouse gases interacts with earth's natural systems. These systems have stayed in balance for over eight hundred thousand years, until human activities such as agriculture, burning of fossil fuels, and deforestation, shattered that balance in astonishing ways.

The earth has a remarkable set of complex, interrelated systems that combine together to regulate its temperature. Solar radiation hits the earth every day (even on cold, wet, Melbourne days when it feels like we have no sun). Much of that radiation is absorbed by the oceans and the land, and then re-radiated out into space at night, although some is immediately reflected, particularly by light coloured surfaces such as ice. This is actually why cities form heat islands that are hotter than the surrounding countryside – because they tend to have many more dark surfaces, such as roads and rooves, that absorb the sunlight.

That heat that is absorbed during the daytime is then re-radiated off into space at night, which is why the earth doesn't simply get hotter and hotter as it absorbs the heat. If this was the whole equation, earth's average temperature would be around -18 degrees celsius. Fortunately for us, and for the plants and animals we depend upon, the greenhouse gases in the atmosphere – mostly carbon dioxide, water, and methane – trap some of that heat, maintaining the earth's average temperature at a much more livable 15 degrees celsius.

This difference of 33 degrees is due to the greenhouse effect. Our lives depend upon it. But this average temperature has been maintained for hundreds of thousands of years by a CO_2 concentration in the atmosphere that, over 800,000 years, has varied between 185 and 278 parts per million (PPM). So for a PPM of, say, 240, this means that 240 of every million molecules of atmosphere are CO_2. That's really small – only 0.024%. This tiny proportion of CO_2 (plus a little water and methane) is enough to change the earth's temperature by 33 degrees celsius! That's 91.4 degrees, for those of you who speak Fahrenheit.

To get a sense of the variation, you can look at the CO_2 measurements taken at the Mauna Loa Observatory in Hawaii since 1958. This is known as the Keeling Curve, because the measurements were started by David Keeling of the Scripps Institution of Oceanography. These measurements are widely regarded by scientists as accurate reflections of global CO_2, because they are taken at a height of 3400m, so they are relatively untainted by local industrial sources of CO_2.'

You can see from this graph that CO_2 levels in 1958 were below 320 ppm. There is minor seasonal variation, but the average is tracking strongly upwards. The most recent measurement at time of writing was 412. Given that CO_2 concentrations have been stable between 185 and 278 ppm for the last several hundred thousand years, that's a terrifyingly sudden increase.

It is even clearer if we include ice core data that shows CO_2 levels trapped in bubbles in the ice. Carbon dioxide can be measured in ice cores because when the ice freezes, it contains bubbles of air that effectively encapsulate the atmosphere at the time.

Weekly CO2 Concentrations Measured at Mauna

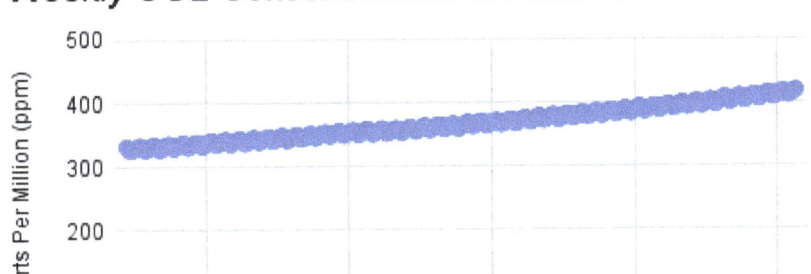

CO2 measurements in the atmosphere taken at Mauna Loa Observatory in Hawaii 48

This data goes back around 400,000 years, and the difference between the natural variation we see at the beginning of the graph, and the sharp rise at the end clearly shows the unprecedented scale of change we are causing.

We have known that increasing CO2 in the atmosphere would increase the Earth's temperature for over 150 years. Eunice Foote conducted rigorous experiments in the 1850s that demonstrated the heating effect of CO2. She even postulated the effect of increased CO2 in the atmosphere on global temperatures. Though her work was presented to the American Association for Advancement of Science in 1856 (by a man), it was not published in the proceedings, and her work was quietly buried, due, as far as we can tell, to rampant sexism. In 1895, Nobel prize winning Swedish chemist Svante Arrhenius presented a paper in which he calculated that doubling the CO2 in the atmosphere would cause a temperature rise of 5-6 degrees celsius50. It's hard to argue that the science is new, or contentious and untested.

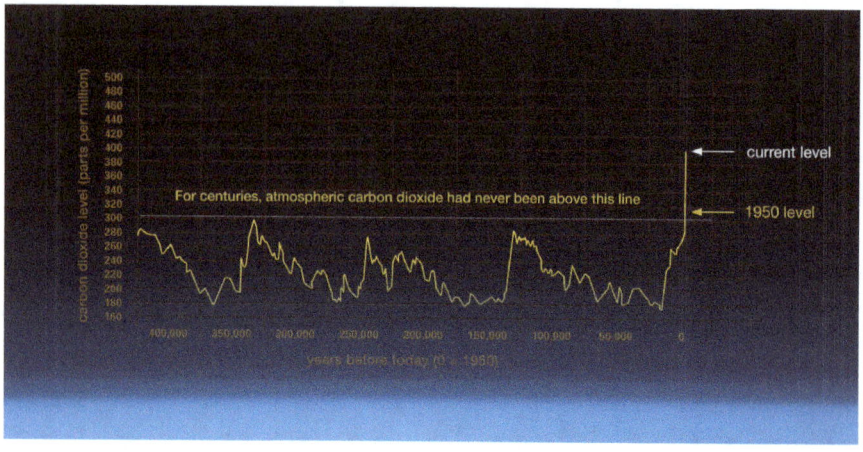

CO2 measurements from ice cores dating back 400,000 years, added to modern empirical data49

Climate Change is clearly not news. Anthropogenic climate change, or the idea that changes in the climate are being largely caused by human activities, is not news. We have been tracking the changes in CO2 levels for decades, and we have known that CO2 was a crucial factor in climate change for well over a hundred years. Arrhenius himself showed that human increases in CO2 would change the climate, though he predicted it to happen over many hundreds of years, given the low rate of human CO2 emissions in his time. As industrialization rapidly pushed up the rate at which we emit CO2, the temperature started to rise faster.

What are the consequences of increasing atmospheric carbon? The IPCC has predicted increasing storm activity – more storms, and more intense storms. We are already seeing that. It has predicted more bushfires, and more intense ones. We are seeing that, too. We will have longer droughts, more intense floods, and mass extinctions. Crops will fail. Oceans will rise. Diseases will spread. A recent report showed that climate change has already destroyed over half of the Great Barrier Reef in the last 25 years.

Action on climate change is often decried as too expensive. Some people claim that fossil fuels are cheaper than renewables, despite the fact that fossil fuels are only cheap because of heavy government subsidies and a complete failure to factor in the so-called externalities such as the impact of particulate pollution from coal and oil on population health[51], the health of the workers who mine, process, and refine fossil fuels, and other environmental impacts.

This looks true up front, because renewables are expensive to set up. It's quite expensive to build a windfarm, install a lot of micro hydro systems, install solar panels or build solar farms, plus there is a significant up front cost in reforming the electricity system so that it can manage the aspects of distributed renewable energy systems that are different from one large, central, generation plant. If you compared that with the cost of building, say, a new gas fired electricity generation plant, though, suddenly renewable energy looks more affordable. Cost comparisons are easy to rig when you compare the cost of generating power from an existing coal fired plant with the cost of building new renewable systems. More to the point, if you consider ongoing costs, renewable energy is vastly cheaper over the long term[52].

We also know that the cost of inaction on climate change will be massive, in the trillions (at a minimum)[53], and that they are already stacking up in the form of extreme bushfires, droughts, and storm events[54]. We only need to look at the devastating 2019/2020 Australian bushfires, where over 17 million hectares of land were burned, or the devastating 2021 floods, which covered many of the same areas, to know that our climate is in trouble.

The science has been clear for more than a lifetime. The logical, rational, evidence based response would have been to start the shift to renewable energy decades ago. Instead, carbon emissions continue to rise, with catastrophic impact on atmospheric CO_2, as the NASA graph above clearly shows.

Which is entirely unremarkable, if you think about it. If you pour a particular gas into a container, that container is going to contain more of that gas. What we struggle to understand, I think, is that the earth's atmosphere is finite. It is, in fact, a container. The things we pour into it – whether they're particulate pollution from industry or bushfires, other toxic by-products of industry, or carbon dioxide – are being poured into a finite space that obeys physical laws. Eunice Foote's experiments with carbon dioxide in a small chamber have direct implications for carbon dioxide in the much larger container that is the earth's atmosphere.

Lobbyists and the fossil fuel industry argue that renewable energy cannot supply all the power we need – and if we wanted to switch overnight, that would of course be true. But many experts and energy industry planners have researched the problem and repeatedly found that 100% renewable energy is an entirely feasible target[55].

Even the Australian government, no friend to climate science (or science in general), ratified the Paris agreement to keep warming below 2 degrees, beyond which catastrophic changes will occur. But to do this we need to rapidly transition all of our power generation to renewable energy, as well as electrifying most of our industry and transport.

The science is in. We need to decarbonise. But instead of building policy based on evidence, somehow climate change and energy policy have become political footballs, and we are hurtling towards disaster. In the 1990s David Suzuki, Canadian scientist and environmentalist phrased it most eloquently:

"It feels as if we're all in a giant car, hurtling towards a brick wall at 100 miles an hour, and we're arguing about where we should sit. There are people screaming 'stop! look out! turn the wheel!' but they're all locked in the trunk."

It's very clear that if energy policy were science and evidence based, we would have fully decarbonised long ago. We have increasingly centralized our power supply into large coal or gas fired power stations requiring high voltage transmission lines (involving considerable losses due to sheer distance) to transmit the power to where it is needed. This has worked against our transition, as renewable energy is by nature easily distributed and local. Leaving the problem to future generations to deal with has left us, the future generations now at the (hah) coalface, with a much harder problem than it would have been if we had started investing in renewable energy decades earlier.

Joel Gilmore, Regulatory Affairs Manager at Infigen Energy, makes the downside of delaying action clear: "We talk about decarbonising Australia – there's a lot of work to do, and we're not doing it now. If you genuinely believe that we'll eventually get on board and act on climate change, the longer you delay acting now, the more pain you have later. If I need to build 100 windfarms over the next 25 years, that's pretty doable. It's only 4 a year. But if I delay 10 years, it's a much bigger problem. Similarly if I have to close one coal powered station every 10 years, that's doable. But if we wait for all coal power stations to get old and die, we literally don't have enough cranes in Australia to build all the wind farms we'd need at that point."

In many countries, including Australia, Climate Change has become weirdly politicised. Rather than considering the evidence and acting on it, acting on the science of climate change seems to be seen as a sign of weakness on the right, and due perhaps to the pressure from right-wing media, it's used as a scare tactic against the left, to the point that no-one from the major parties is brave enough to take action. We know, without doubt, that our failure to act will have catastrophic consequences, yet the short term political challenges outweigh our very future.

An energy industry expert who I interviewed for this section told me: *"The economics stack up. The technological things stack up. If you coordinate it, if someone gets it done, it absolutely can be done. The government needs to listen to expert advice and agree on a coordinated plan, rather than constantly making captain's calls, which is what we're seeing. Electrons don't understand political boundaries."* Unfortunately, because this issue is so politically charged, I am only able to use that quote anonymously. It is absurd that the idea of listening to expert advice is something that is dangerous to suggest in public.

An evidence based shift to clean, carbon free energy sources is a clear win. Quite apart from the impact on climate change, it involves less pollution, cheaper long term energy, and more jobs, so an evidence based policy approach would have to prioritise a shift to renewable energy. Yet here we are, in 2020, with the latest Federal Budget in Australia prioritising fossil fuels rather than renewables, and Australia's Clean Energy Fund and the Australian Renewable Energy Agency being pushed to fund gas development rather than focusing on the development of carbon free alternatives such as technologies based on hydrogen and ammonia[56].

Evidence based climate policy would mean intense investment in renewables, immediate action to decarbonise, a global halt to deforestation, and a vast effort to reforest, revegetate, and rehabilitate our environment. Rather than hippy pipe dreams, these are just some of the actions that it is crucial for us to take in order to reverse the catastrophic consequences of increasing atmospheric carbon.

Evidence Based Welfare

There is a lot of debate and polarised rhetoric around approaches to homelessness and unemployment. Much depends on the intent of Government programmes – is the intent to ensure that everyone has what they need, or is it to ensure that no-one gets anything they are not entitled to? As the author of Automating Inequality, Virginia Eubanks, notes[57], these two approaches lead to very different outcomes.

So let's start from a dispassionate, purely economic perspective. Multiple studies have now confirmed that it is cheaper to simply provide housing than to leave a person sleeping rough[58]. This is because there are two main costs of people sleeping rough. One is the cost of policing their non-violent rule breaking – moving them on from places they shouldn't be in, penalising them for sleeping on the streets when they have nowhere else to go, etc. The other is the health costs of sleeping rough, which then become societal costs when homeless people present to emergency rooms repeatedly for issues related to exposure and stress that would not happen if those people had safe places to stay.

We also know that trickle down economics doesn't work, so the idea of giving more money to companies and rich people does not result in a flow of money out into the economy, but giving money to poor people goes directly back out into the economy in the form of purchased goods and services[59].

Yet, as I write this, in late 2020, the Australian Federal Budget has just been released, promoting tax cuts for the rich as a way to stimulate the economy. The trouble is that we know this approach does not work. The most effective way to stimulate the economy is to give money to poor people, because they will spend it. It's important to note that, contrary to popular rhetoric, they will mostly spend it on things they need – food, housing, healthcare, education, clothing, and the like – rather than on drugs and alcohol.

According to a recent study by The Australia Institute[60], higher taxes actually correlate with stronger economies, greater wellbeing, and pretty much any positive measure you care to name. The definition of a progressive tax system is one in which income tax rises with incomes – so higher income earners pay higher tax rates. This reduces inequality, and maximises the government's ability to provide essential social services such as health and education, as well as essential infrastructure spending on things like electricity networks, public transport, and high speed internet.

You can see why it's called a *progressive* tax system – because it generates forward positive progress for the whole of society.

By contrast a regressive tax system is one which takes the same amount of tax from high income and low income earners alike. Australia's Goods and Services Tax is a regressive tax, because the tax is the same rate per item or per service for everyone, rather than a different rate according to your income.

This, of course, highlights a fundamental ideological divide. If your goal is to create a system where inequality is minimised and essential services are freely available to everyone, then a progressive tax system is clearly the way to go. If, however, your primary goal is to enable rich people to get richer without limit, then regressive taxes are perfectly applicable.

There is also a severe ideological divide around welfare. There are two main ways of characterising the welfare system:

1. An artefact of the nanny state that is open to abuse by slackers looking to get something from nothing, or
2. A way of ensuring that everyone is looked after, whatever their circumstances.

At the heart is this divide is a fundamental difference in the view of humanity. Are we, deep down, people who want to work, contribute to society, and feel good about ourselves, or are we, in fact, people who always want to get something for nothing. Fundamentally lazy. Rotten with sin.

But poverty is not a lack of character. Poverty is a lack of cash. The Australian Council of Social Services studied the Australian Government's preferred welfare system, the Indue cashless debit card[61]. As the report states: ""The cashless debit card quarantines 80% of a person's income support payment to a debit card which cannot be used to purchase alcohol, gambling or withdraw cash, in an effort to change behaviour associated with addiction."

The government is so committed to this card that, despite it costing $10,000 per person to implement, they are determined to roll it out nationwide. And yet the report notes: "There is no reliable evidence that restricting access to cash and banning the purchase of alcohol or gambling helps people receiving income support or their communities."

The media loves to highlight stories that embody the "people are born dreadful and lazy and cannot be trusted" narrative, because outrage sells. But Rutger Bregman argues differently. In his book *Humankind: A Hopeful History*, he lays out considerable evidence that our sociability and our ability to collaborate are actually the reasons why humankind has been so successful.

This dovetails neatly with the evidence from every society that has trialled a Universal Basic Income (UBI) – where every adult gets paid a living wage by the government. A UBI means enough money to reliably have housing, health, food, and education covered, regardless of how much they earn or how much they work. Trials of UBIs in Canada and Finland, among others, have actually seen no significant change in the amount people work. In fact, the data shows that only two groups of people engage in less paid work when they receive a UBI – new mothers, and students[62].

In addition to funding everyone's basic needs, we also know that education is the key to lifting people out of poverty.[63] The increasing inequality between public and private schools in wealthy countries such as Australia is another factor in maintaining the structural inequality that holds us back. In my school visits I've seen many public schools with rundown buildings, peeling paint, broken air conditioning and limited heating. The students learning at these schools must overcome all of the environmental challenges before even beginning to learn. (And that's not taking into account how many of these students are likely to be going hungry.)

At one High School I visited, the teachers told me that some of their students were exhausted in class, because they work long hours after school as the sole income earners for their families.

I've also visited private schools with advanced gymnasiums, theatres with high tech, retractable seating, swimming pools, tennis courts, and the latest technology. Students at these schools are well supported by their environments, and often have nothing to do but relax and learn.

We also know that postcode (a crude proxy for Socio Economic Status) is a disturbingly reliable predictor of academic outcomes[64]. Students living in poverty have lower academic results. Lifting families, and indeed whole areas, out of poverty, is likely to close this gap.

As noted in the Gonski Report[65], Australia's review of educational funding completed in 2011, *"Funding for schooling must not be seen simply as a financial matter. Rather, it is about investing to strengthen and secure Australia's future. Investment and high expectations must go hand in hand. Every school must be appropriately resourced to support every child and every teacher must expect the most from every child."* Though the Gonski report received bipartisan support during the subsequent election campaign, after the Liberal/National Party coalition won the election many of the recommendations were quietly scrapped.

We could fund universal basic income, universal healthcare, and free high quality education for all, by implementing a truly progressive taxation system that expects higher income earners to contribute more to society. Given that we know that countries with higher tax rates have stronger economies, that providing for people's basic needs is cheaper than dealing with the mess that results when people are homeless and destitute, and that a high quality education provides a path out of poverty, it is difficult to understand why our political and economic policies don't match the evidence.

Evidence based education

We know that one of the attributes of highly effective teachers is extensive knowledge of their disciplines. This is one of the reasons I was profoundly shocked to discover, when I first became a high school teacher, that one of the consequences of a failure to invest adequately in education is that in Australia, a teacher can, and likely will, be asked to teach subjects for which they are entirely unqualified. This is known as Out of Field Teaching. A 2015 report from the Australian Council of Educational Research found that around 26% of teachers at year 7-10 level were teaching subjects they were not qualified to teach[66]. As noted in the report, this is of huge concern.

Out of field teachers lack not only domain knowledge, or knowledge of the topic they are expected to teach, but also domain specific teaching knowledge, known as Pedagogical Content Knowledge, or PCK. It is the awareness of where students can get stuck, which things are difficult to learn, and a range of different ways to effectively convey complex ideas.

Because I was new to teaching and had only experienced the Australian system, I assumed this was an artefact of an intractable problem – that the number of teachers, the subjects those teachers were qualified to teach, and the number of classes that needed teaching was simply impossible to match precisely. If that is the case, you would assume that all countries around the world face the same issue. However, it is not the case in Germany, Spain, or France. Teachers in those countries simply could not be asked to teach subjects they are not qualified to teach. It just doesn't happen. It *is* the case, though, in England, the United States, and Australia.

Out of field teaching is not inevitable. If other countries can solve it, the problem is clearly not intractable. It is difficult to know why it is accepted practice, although in some areas, such as computing, there is a clear shortage of qualified teachers – largely, I suspect, because anyone with computing skills can earn far more money in almost any other industry.

Evidence based Computing Education

When I started studying Computer Science at university in 1989, the first year programming subject used a programming language designed specifically for teaching, on the premise that it would be easier to learn than a programming language designed for experienced programmers. Given many of the languages around at the time, which were notoriously difficult to program in, let alone to learn, this seemed a logical move. Languages such as C, Lisp, and Prolog were widely used as introductory languages at the time – and these languages are definitely not designed with learning, or usability, in mind!

When I started my PhD in Computer Science Education, I took this idea and ran with it, designing a new teaching language: GRAIL (GRAIL, in a startling display of hubris, stood for Genuinely Readable And Intuitive Language). Which is rather ironic in hindsight, because the one thing I learned for sure in my PhD was that a teaching language was one of the worst possible ways to teach programming.

Why is using a teaching language such a bad idea? Because one of the most crucial factors in learning anything is intrinsic motivation[67] – ie the motivation that comes from doing something fundamentally meaningful, as opposed to extrinsic motivation, which is where marks and other external rewards are used to try to motivate student behaviour.

Using a teaching language removes the motivation of learning something that has a real purpose – something useful, that can be applied. If you are learning programming, it is almost certainly so that you can use it to *do* something – whether it's modeling the spread of a virus, building a website, or finding out what a dataset can tell you. Using a teaching language effectively divorces learning from purpose, making the goal of doing something real a little further away, a little less reachable. Particularly when the sample problems used for teaching are also toy problems with no real world application. Now students are using a language that they can't use for anything real, to solve a problem that is not meaningful! After a few years battling through with the teaching language, the University switched to C++ – an awful language for learning, but with motivation baked in, because C++ was being used for *a lot* of practical applications at the time.

Had the switch to a teaching language (and back) been carefully documented and evaluated, things might have been different. Unfortunately, the decision on which language to use for introductory programming language tends to be made on the basis of ideology rather than evidence, which might be why the same university switched to using a block-based programming language – a language designed for teaching programming to kids – in first year in 2012.

There is a huge number of papers on motivation and programming. Unfortunately, they are mostly of the form "We taught a thing. The students LOVED it. They were SO motivated!" with no objective, systematic evaluation of before and after scenarios.

While we're talking about evaluation, consider the immense proliferation of Women and Girls in STEM programs, designed to encourage more girls to study and work in STEM. A recent study by Merryn McKinnon from the Australian Science Media Centre found that, of 337 women in STEM programs or initiatives studied, only 7 had *any* publicly available evaluation, of which only a single program went beyond asking the students whether they liked it[68].

Without properly evaluating the outcomes, how can we be sure of what we are achieving?

Teaching the teachers

It's a common misconception that once you have learned programming (or anything, really), you can teach it. This assumes that knowing something is the same as being able to teach it. Unfortunately, the people who are best at a skill or content area are often some of the least able to teach it – because they find it very difficult to understand why anyone else would find it difficult. This is a significant part of the "Pedagogical Content Knowledge" of computing. PCK is something that can be learned, but it definitely does not come prepackaged with any skill. A significant part of the problem here is that most teachers did not do any programming or computing of any sort as part of their own schooling. Even people who dropped maths and science as soon as they were allowed to will have done both subjects until year 10 level. Until recently, it was possible to go through your entire school career without doing any computing more complex than using Microsoft Word.

One key problem with the "learn to program and then immediately teach programming" approach is that it is exceptionally difficult to identify where a program is going wrong, or debug it, without significant experience in debugging code. I can usually look at a student's code and identify problems very quickly because I have been programming for over 30 years. Someone who is learning, or has just learned, does not have those skills.

Programming is widely taught now, despite the lack of qualified teachers. This has implications for the quality of the education, and the amount of support that students receive. We know that girls' confidence in STEM skills is lower than boys', even when their actual abilities are the same. We also know that girls' anxiety in maths is exacerbated by teachers with maths anxiety, which is, sadly, a common scenario in primary schools[69]. It seems likely that the same will apply to computing education: there is a formal requirement in Australia for students to learn programming and computational thinking from Primary School, but most current primary school teachers have not been trained to teach computing at all, and many find it daunting, to say the least. Without significant support for building teacher confidence, student confidence in Computing is likely to suffer an even worse fate than maths – because at least primary teachers have been taught to teach maths (and they have studied it at school for at least 10 years)!

I did a quick survey of Primary Teaching courses available currently in Australia. Swinburne University offers a Primary Teaching degree with a single unit that mentioned technology[70]. This is the "Teaching Science and Technology Unit", and the content is listed as:

- Designing Science lessons
- Being inclusive and safe in Science
- Cross-disciplinary and single subject approaches
- Teaching strategies for Science
- Developing Science lessons based on National curriculum
- Assessing Science
- Integrating and assessing Science, Design, and Digital tech

It is interesting to note that there is nothing there about developing Digital Technologies lessons based on the National Curriculum, even though the Digital Technologies Curriculum was adopted in Victoria in 2017. (In contrast, there are 3 separate units on teaching Maths.)

The University of New England Bachelor of Education (K-6 Teaching) also offers one unit in its core curriculum on teaching "Science and Technology"[71]. The Curtin University Bachelor of Education (Primary Education) contains no digital technologies units in its core degree, though there are some electives, including one on coding[72]. But it is possible – indeed, apparently highly likely – to become a qualified Primary Teacher in 2020 without ever learning how to teach Digital Technologies, despite it being a core part of the Australian Curriculum.

The year I left, our school began a literacy programme that was delivered by teachers on the condition that the program had to be delivered exactly as written – the same slides, the same talking points, the same activities, and the same workbooks. The activities were the same for all teachers, rather than tailored to particular teaching methods, which led to activities where, for example, maths teachers had to come up with examples of how to address certain literacy issues in an English class. Rather difficult to see how this would be motivating. There were spelling and grammatical errors on the slides. It was an astonishingly poor way to communicate, let alone to teach.

In Australia, teachers have to do compulsory professional development in order to remain registered as teachers and allowed to continue teaching. It was extremely common, when I was teaching in a secondary school (I left at the end of 2017), for the PD delivered to teachers to flout the most basic rules of education, just like the literacy program did. We know, for example, that "chalk and talk" or standing up the front lecturing is one of the least effective teaching methods in use today. An extraordinary amount of teacher PD is death by Powerpoint – one person reading Powerpoint slides, or talking through an endless slideshow. This is the example we set our teachers.

If we zoom out for a moment and examine education more broadly, it is strange to note that there is a near universal reliance on exams as a primary form of assessment. Some subjects here and there use project or portfolio based assessment, but by far the most common way of assessing students – whether for entry into school or university, or for mastery of a subject – remains a written exam.

My 17 year old daughter, Zoe, is about to complete her final school exams. In Victoria, those final exams are combined with school based assessments (mostly written tests) to create the Australian Tertiary Admissions Rank: Not a score, but a ranking used to determine university entrance. The exam for English involves responses to prompts that require analysis of set texts. Points are given for use of quotes, but no notes or books are allowed into the exam, other than a dictionary. This means that students who happen to be good at memorising quotes will get a better mark than students who are not. It's a test of memory, not of reasoning, understanding, or analytical skills. As Zoe commented: *"This method of testing people's English skills is not testing their English skills. It's testing how well they can bluff under pressure, and WE DON'T NEED MORE OF THAT!!!"*

Closed book exams often rely heavily on recall of facts and figures, dates and definitions, that are not tests of skill or understanding, but tests of memory.

The obvious advantage of exams is that you can have your whole cohort sit the same exam, at the same time, under the same conditions, and you can even blind mark the exams by ensuring students identify their papers with student ID rather than name. This, on the surface, looks like a fair, objective, and effective way to check student learning.

Unfortunately, closed book exams that require memorization encourage cramming, where students learn as much as possible in the time immediately preceding the exam, and promptly forget it as they cram for the next subject.

The rapid onset of home based learning and assessment due to covid19 has also severely challenged the reliance on exams. They have either had to be ditched altogether, or conducted online. As soon as students are sitting an exam remotely, it effectively becomes open book, as there is no reliable way of checking that students are not accessing other resources.

Also, of course, student performance in exams is a wonderful predictor, for the most part, of student performance in exams. Some students suffer from exam-induced anxiety, others don't work well under the extreme time pressure of exams. Some students (myself included) struggle to write sufficiently neatly to be confident of being understood by the examiner. And some students interpret the questions in a different way, and without being able to test their interpretation (most exam conditions don't allow this), they may fail the exam even while demonstrating a high level of competence in the subject material.

It is extraordinarily difficult to write an error free exam. I've written many questions that were open to multiple interpretations. I once wrote a question for my year 11 Information Technology students that technically merited full marks if they left the question blank. (For programmers, I defined a function in Python but didn't call it, and asked students to write down the exact output of the program – which had no output. I was looking for the answer "there is no output, the program will not run," but a blank page was a perfectly correct answer.) It is, of course, possible to write exams that are much better than average, but, as with so many aspects of teaching, there is rarely time to do the job properly. Even final year 12 statewide exams, which are rigorously tested, have been known to have errors, confusing questions, or material that is not actually covered by the course.

What if we assessed students on their problem solving, creativity, and project work? On their ability to collaborate, to evaluate and challenge their own work, and on the processes they use to test their own answers? What if, in short, we tested them on the skills we say we want them to come out with? In Chapter 4 we will see that the graduate attributes specified as outcomes for every university surveyed include problem solving, creativity, and ethical behaviour, all of which are very difficult to test effectively in a closed book exam.

It's clear that in education, too, we are not following the evidence, investing in what we know works, and implementing effective strategies to solve the problems that we have. It's not that we don't know how. It's that – like in medicine, climate science, and welfare – we can't seem to get past ego, ideology, and stubbornness in order to make progress.

How do we fix it all?

We've examined four areas in this chapter – Medicine, Climate Change, Welfare, and Education – and it's clear that ideology all too often plays a greater part in our political decision making than science and evidence. This is one of the key reasons why it's so crucially important to raise our children, from the moment they enter formal education (and before!) to be rational heretics. To solve challenging problems, and then rigorously test their solutions. To ask difficult questions and objectively analyse the answers. To understand data, and how it can be used both for and against us. In the next chapter, we'll look at science education, and ask whether it is teaching science, or confirmation bias.

3

Science is Solved

A History of Heresy in Science

Much of our Science, Technology, Engineering, and Maths (STEM) education starts from a foundation of facts and known answers. This teaches our kids that the point of STEM is to Get The Right Answer, whereas the actual point of real world STEM disciplines is to fix things, understand things, and solve problems.

First of all, let's define our terms. Heresy is an opinion profoundly at odds with what is generally accepted. And yet, heresy has been crucial to our scientific development. In the 1840s Ignaz Semmelweis came up with the radical heresy that doctors washing their hands before (and after) surgeries prevented disease. Prior to this doctors went from autopsies to childbirth without washing their hands or changing their clothes. And they wondered why people died. The idea that this could cause disease was considered so ludicrous that it took decades for the idea of washing hands to be accepted. Semmelweis was so ridiculed and pilloried that his colleagues committed him to an asylum where he was beaten and died.

In another well known heresy, Galileo Gallilei so outraged the church with the idea that the earth revolves around the sun, rather than the other way around, that he was accused of literal heresy and committed to house arrest. He only narrowly escaped death.

In 1912 Alfred Wegener began to publicly advocate the idea that the continents moved over time – what became known as continental drift. He, too, was widely ridiculed, and did not live to see his ideas finally vindicated.

In 1917 Alice C Evans made the laughably heretical suggestion that milk should be heated to a high temperature, or pasteurised, to kill bacteria that could be harmful to humans. She was not taken seriously, being a woman and without a PhD (which, by the way, were not offered to women at the time), and it took over a decade before milk was regularly pasteurized in the US. After her discovery but before its general acceptance, Alice became significantly ill with Undulant fever, a disease caused by one of the bacteria found in raw milk.

In the 1940s and 50s, Barbara McClintock discovered that genes aren't static sets of instructions passed from generation to generation, but that they can be regulated – turned on and off – by other parts of the genome. She described the reaction to this discovery as "puzzlement, even hostility", but in the end her research radically changed our understanding of genetics.

In the 1960s, Frances Kelsey of the American Food and Drug Administration refused to approve Thalidomide for use as a morning sickness drug, because she was concerned about the lack of data about whether the drug could cross the placenta, and directly affect babies' development in the womb. This averted thousands of birth defects in American babies. Sadly, other countries were not so cautious[73].

More recently, Marshall and Warren's original paper on ulcers being caused by bacteria rather than stress was rejected and consigned to the bottom 10% of submissions. Barry Marshall eventually drank *helicobacter pylorii* – the bacteria that causes ulcers – to prove it, thus inducing an ulcer which he then cured with antibiotics.

And if you want a really recent example, let's talk about Linsey Marr, an Aerosol Scientist from Virginia Tech who also studies infectious diseases[74]. In 2011, Marr tried to publish her findings from a study she conducted that sampled the air in various public spaces. Those samples found the flu virus where doctors said it couldn't be – suspended in the air. At that time the accepted orthodoxy was that flu was transmitted by droplets, which do not remain airborne. In other words, to catch the flu you need to pick up an infected droplet from contaminated surfaces, or be in the direct line of someone's cough or sneeze. Marr proved that there was actually enough flu virus suspended in the air to give people the flu. The paper was rejected as wildly implausible, because the accepted orthodoxy was that only particles of 5 microns or smaller could remain airborne. Marr persisted, and when the WHO insisted that covid19 was not transmitted by aerosol particles, she and many others beat their heads against the brick wall of this entrenched 5 micron theory. Eventually it was proven that covid19 is actually primarily transmitted by aerosol particles, and Marr's work was vindicated – but not before hundreds of thousands of people died unnecessarily, due to the WHO's wrong advice.

The kicker to this story is that the rigid 5 micron idea came from a study of tuberculosis in the 1930s, which showed that only particles smaller than 5 microns in the atmosphere could actually lead to tuberculosis infections. The study was accurate, but misinterpreted, because tuberculosis needs to lodge deep in the lungs to cause infection, and only tiny particles (smaller than 5 microns) can make it that deep into the lungs. But most viruses don't need to go anywhere near that deep in order to cause infection, which means that much larger particles can be dangerous.

So the original study didn't even show that only particles smaller than 5 microns could remain in the air. It showed that only particles smaller than 5 microns could transmit tuberculosis. But that "fact" was rigidly applied in medical science for nearly a hundred years, until Marr and her fellow heretics successfully challenged it. And though they've won the battle with respect to covid, I bet many other diseases are still being treated according to this "fact".

It seems like heresy is a pretty dangerous business.

In fact, a lot of scientific breakthroughs have been considered heretical. Especially in medicine. Scientific progress is achieved by continually testing our understanding, and trying to prove our theories wrong. It actually slows down the progress of science if we treat theories as facts, rather than theories, and become attached to them. It leaves us unwilling to see them challenged. So how do we change our approach to science to ensure that rational, ethical scepticism becomes the norm? If science is not facts and known processes, what actually is it?

What is science?

The common perception of science is that it is a discipline of facts and right answers. This is absolutely the way we teach it, but it's the opposite of what science is. Science is a way of exploring and understanding the world, and of solving problems. By its very nature science deals with uncertainty, and constantly proves itself wrong as new information becomes available.

Scientific theories are based on the information we have right now. Sometimes we can't see, measure, or understand enough to explain a phenomenon fully, but we have a model we think is right, and it's right enough to help us understand some parts of the way the world behaves.

For example, before Galileo, people used to believe in Aristotle's idea that everything revolved around the Earth. It's an idea that makes sense from our reference point. We can see that the sun goes across the sky. So does the moon. We can't feel ourselves turning. The idea that the Earth is a giant ball spinning us around is something we can't directly observe and that, frankly, makes no intuitive sense. Also, we are quite obviously the most important thing in the world (Christianity told us so!), so it's only logical that everything revolves around us.

But it turns out that when you use a telescope you can observe the planets and find that they do things that don't make sense if everything is revolving around us. For example, one of the things that Galileo noticed was that Jupiter had moons revolving around it. Which was absurd, because nothing was supposed to revolve around anyone but us. It took the world quite some time to come to terms with the idea that the Earth was not the centre of the universe. Indeed, the very idea was so outrageous to the church, it nearly cost Galileo his life.

The thing is, Aristotle wasn't wrong, for his time. He had a theory which adequately explained the universe as far as he was able to observe it. There was no sign that the Earth was moving – no feeling of motion, none of the consistent wind you get when you're moving at any speed, no unexpected oddities in the motion of falling objects around him. His theory fit the observations that he had available to him. Which means that Galileo's discovery was science at its best. "Here are some new observations that mean our old theories were wrong. What new hypotheses can we come up with to explain them?"

The nature of science is that we constantly test our theories with the best data we have available, knowing that new data might show that some of our theories aren't good enough, and some are even outright wrong.

Underpinning the way we do (or should do) science is the concept of fallibilism: starting from the assumption that your ideas could be wrong.

Fallibilism requires that you work to disprove your ideas, rather than trying to prove them right. It's the idea that we can't accept a hypothesis as even plausible until we have tried our hardest to prove it wrong, without success. In this model a theory can never be proven, only disproven, or supported by the available evidence. A "supported by the available evidence" theory, of course, is not nearly so satisfying as a proven theory. It leaves a whole lot of uncertainty, which is disconcerting and uncomfortable. But knowing that a theory can't be proven changes your whole approach. When you start from the position that a theory can be disproven or supported, you must be open to believing that new evidence might come to light that disproves it, or at least shows it to be inadequate.

Fallibilism is a crucial antidote to our human tendency to become attached to our own ideas. We want our ideas, beloved fruit of our labouring brains, to be right. Sometimes we are so invested in their rightness that we allow confirmation bias to persuade us that they are. We look for the evidence that supports our ideas, and we ignore or discount the evidence that counts against them.

Fallibilism is sadly not our default approach. To a great extent, our academic structures work against it, because academics are actively judged and measured by the number of publications they put out. There's no space or reward in that system for rigorously testing and critiquing your results, only for getting them into publication quickly.

We can see examples of this everywhere, from academic publishing to popular science writing. In his book *Humankind: A Hopeful History*, historian Rutger Bregman lays out the hypothesis that humanity is fundamentally prosocial and altruistic, rather than antisocial and selfish. To support his ideas he explores a range of famous case studies – from the history of Easter Island to the Stanford Prison experiment. It is, indeed, a hopeful history, but his coverage of the science behind the famous case studies is actually rather demoralising, because it is a tale of either outright scientific fraud or a desperate failure of fallibilism. In each case Bregman lays out the famous, media-worthy version of the studies, and then painstakingly dissects the evidence to show an entirely different scenario.

For example, in Jared Diamond's discussion of Easter Island in "Collapse: How Societies Choose to Fail or Succeed," Diamond talks about how the islanders cut down all of the trees in a frenzy of greed and a kind of statuary arms race. He describes how the loss of trees led to the collapse of farming productivity, and even turned the islanders into cannibals. The same evidence, though, turns out to support quite a different hypothesis.

As Bregman writes: "For the Easter Islanders, deforestation was not that big a deal, because every felled tree also freed up arable land. In a 2013 article, archaeologist Mara Mulrooney demonstrated that food production actually went up after the trees were gone, thanks to the islanders' use of savvy farming techniques like layering small stones to protect crops from wind and retain heat and moisture." Biologists also suspect that the introduction of rats, who fed on the seeds, was a large factor in the demise of the trees[75]. Jared Diamond even notes that every seed found was gnawed on by rats, and hence would never germinate. Nonetheless, he ascribes the bulk of the deforestation to human activity.

He also builds much on an intense rivalry between different tribes on the island that appears to be entirely imaginary. "We know of not more than a hundred pukao, reserved for statues on the biggest and richest ahu built late in Easter prehistory. I cannot resist the thought that they were produced as a show of one-upsmanship. They seem to proclaim: "all right you can erect a statue 30 feet high, but look at me: I can put this 12 ton pukao on top of my statue. You try to top that, you wimp!" He goes on to compare the statues and Pukao to Hollywood mansions. He might not have been able to resist the thought, but he could certainly have tested it more thoroughly.

Diamond refers directly to the clear evidence of integration among all of the tribes on Easter island, as demonstrated by the even distribution around the island of resources that can only be found in one area, such as obsidian, and the stone for the statues, as well as tangible evidence of intensive farming on the uplands, the crops from which were distributed around the island. This seems at odds with the thesis that rivalry effectively destroyed the island. And he fails to mention the existence of an alternate, almost completely opposite hypothesis: that the statues were actually community building exercises designed to bring the entire island together.

Whether Diamond's hypothesis was right or not, there was clearly evidence available when he wrote his book that puts his ideas in question. In a basic academic error he quotes James Cook without ever following the quote to its source to check its veracity. He says that Cook describes the islanders as *"small, lean, timid, and miserable"*.

However, when Bregman checked Cook's notes he found nothing of the kind. "Instead, Cook reports that the inhabitants were *'brisk and active, have good features, and not disagreeable countenances; are friendly and hospitable to strangers."* Hardly the description you would expect of a famine ravaged nation that had turned to cannibalism to survive.

The erroneous Cook quote does appear in other work. The first reference Bregman could find was in a book by Thomas Heyerdahl. So while Jared Diamond didn't make it up, he did commit the fairly basic academic error of not verifying his sources. If you are going to build a significant case on a quote, it is fundamental to effective scientific scepticism, and rigorous research, to track that quote back to its source, to be sure it is accurate.

You might say that this case is trivial. Perhaps nothing vital hinges on whether Diamond's depiction of Easter Islanders as greedy, violent perpetrators of ecocide is accurate or not. But it does illustrate the problems of being too eager to verify your pet theory. And sometimes those pet theories do significant damage.

Take the case of Jennifer Connellan & Simon Baron Cohen's famous gender difference experiment[76]. Connellan showed 1 or 2 day old babies her face, or a mobile with scrambled pictures of pieces of a face pasted onto it and measured how long they looked at it. The study was widely publicised as finding that babies as young as one or two days old showed gender-based preferences for faces versus things. Boy babies, (budding engineers and builders all) showed more interest in the mobile, while girl babies (nurturing nurses and teachers in the making) were all more interested in Connellan's face.

That study is still frequently cited, and often used as evidence that girls are simply less interested in STEM, especially the 'hard' STEM careers that are typically thought of as male, such as Engineering and Computer Science, because boys are "systematic" and girls are "empathic". It's hard wired. Nature, not nurture. Which means we don't need to work to increase diversity – that would simply be 'pushing girls into careers they are not suited to' (not to mention all of the boys who are not 'allowed' to prefer nurturing, people-oriented careers).

Unfortunately, it turns out that there were significant issues with the way the experiment was conducted. Firstly, the faces & mobiles were not presented at the same time, so the difference was recorded as the amount of time the babies looked at the different things, rather than a choice between them, which goes against the standard experimental approach. It's not easy to determine exactly where a baby is looking at the best of times. Secondly Connellan knew the sex of the babies she was studying, leaving a lot of room for confirmation bias, or seeing what she expected to see. Thirdly, and decisively, the study has never been replicated. There are many other issues with the methodology that I won't go into here, but a study that has not been replicated cannot reasonably be claimed to be definitive.

Fallibilism demands that the study be rigorously tested and replicated, and the methodological issues addressed. For instance, it would not be difficult to present the babies to the experimenter in neutral clothing, and with initials rather than first names, so that the experimenter does not know the gender in advance.

In the documentary "No more boys and girls: can our kids go gender free?" the makers conducted an experiment where girl babies were dressed as boys and given male names, and boy babies were dressed as girls and given girl names. Adults were brought in to play with them, and without prompting those adults gave the babies dressed as girls, who were actually boys, dolls and teddies, and the babies dressed as boys, who were actually girls, robots and puzzles[77]. The adults all said the babies were clearly more interested in those things. It would be fascinating to see how the results of Connellan's study fell out if it were repeated under conditions like those!

The Connellan experiment, at this point, looks very like another failure of fallibilism.

Unfortunately the academic model of science actively discourages fallibilism. There are no points for carefully testing & trying to demolish your results or theory, but there are many points for fast and prolific publishing. The number of papers you publish as an academic is a direct measure of your value to your institution, since funding models are tied to what we call research output, but which actually boils down to number of publications. Some funding models contain a kind of ranking system that measures where you publish (ie how prestigious the conference or journal is), as well as how much, but the emphasis is still on pumping out results fast.

The peer review process is supposed to deal with this, by having each paper reviewed by others in your field. Peer review, though, is fundamentally flawed. Reviewers can't replicate the work to see if it stands up – no-one has that kind of time, and papers frequently don't have enough detail to make it possible. They can assess the quality of your writing, but they're not necessarily in a position to assess what you did, without being in the lab and peering over your shoulder the whole time.

This culture and incentive system leads to scientists being deeply invested in their own ideas – because their very careers and livelihoods depend on them. Not on testing them. Just on publishing them. And it's worse than that, because there is no recognition in the funding model for effective public communication of science. Academics are rewarded for publishing in their fields, but effectively penalised by spending time communicating their results to the rest of the world (because that is time spent not publishing in academic settings, or scrabbling for funding).

Some fields are already undergoing a reproducibility crisis, as long-established experiments (like the marshmallow experiment with kids, which is much cited as evidence that some kids have willpower, some don't, and it doesn't change[78]) are found to be irreproducible, and thus extremely dubious. My expectation is that most fields would go through this kind of crisis, if only anyone had the time and the funding to try to reproduce all of the experiments.

What we teach – Science curricula around the world

Unfortunately it's no surprise that not all scientists are committed fallibilists. In addition to the systemic issues noted above that push scientists towards fast publication rather than thorough testing, the very way we usually teach science doesn't tend to support fallibilism. Or, even if our teaching does try to do that, the way we assess it certainly doesn't.

I'll talk more about this problem of assessment in chapter 4, but for now let's consider the Australian curriculum for primary school science. Intended for kids from roughly 5 to 12 years of age, the science curriculum has sections for chemical, biological, physical, and earth and space sciences, as well as a science inquiry skills section.

This curriculum has a heavy emphasis on facts and known processes. It claims to encourage students to explore & research topics, and design experiments to explore problems, but it is primarily a list of facts and concepts that kids are expected to know & understand.

You might argue that with young kids, giving them bite sized chunks of knowledge that they can digest easily is more important than a broader understanding of the nature of science. That exposing them to the complex way our understanding changes that characterises true science is only going to be confusing. But we know that the early years of schooling are when kids' interest in science is either nurtured or crushed. And presenting science as an array of things we know is hardly enticing. Kids love mysteries. They love to feel purposeful. They want to find things out, not be told things.

What if we presented some observations – or, better yet, got the kids to make observations! – and then asked them to come up with hypotheses that explain those observations.

And then get them to test those hypotheses.

And then decide which hypothesis is more plausible given the evidence.

What if we presented them with two hypotheses – perhaps historic ones – and ask them what experiments they would need to perform to test those hypotheses?

In other words, what if we had those kids *doing* science, rather than learning facts?

What if, instead of having kids look for evidence that supports the theories we teach them, we have them look for evidence that contradicts them? The English science curriculum for Years 1 and 2 has a strong emphasis on nutrition, which is one of the areas of science that is notorious for constantly changing advice. Unfortunately it lists plenty of facts and concepts, but no doubt or reference to changes in understanding.

Eggs, for example, were decried for a long time as a source of cholesterol, which was bad. Now, however, we know that there is "good" cholesterol and "bad" cholesterol, and eggs are chock full of the good kind (though I wonder how long it will be before this, too, turns out to be a misleading oversimplification). Exploring the science of nutrition could mean exploring the changes in nutritional advice over the years, and what kinds of science caused those changes.

It's not unreasonable to expect that the UK and Australian science curricula would be similar. What about the rest of the world? I used the Trends in International Mathematics and Science Study (TIMSS) 2015 encyclopedia to explore the science curricula across a range of different countries[79]. Of the 70 countries covered by TIMMS, I explored a cross section of cultures and societies, including Israel, Finland, Lebanon, Russian Federation, South Africa,Italy, Singapore, Japan, France, Morocco, Chile, and the United States. In each curriculum summary I looked for references to five key topics:

- Problem Solving
- Critical Thinking
- Scientific Thinking
- Scientific Method
- Ethics

First, a few caveats: TIMSS does not show the entire curriculum. It is the result of compiling responses to a 2019 questionnaire sent out to countries asking them to describe the priorities of their science and maths curricula for primary and lower secondary year levels. It's not intended to be a comprehensive listing of everything covered, but it is a reasonable guide to each country's priorities in the science they teach. In some cases the TIMSS responses focused exclusively on grades 4 and 8, so it's also possible the ideas I was looking for are covered elsewhere.given the fundamental importance of these ideas, though, one would expect them to be included every year to some degree.

The TIMSS website explains the process: "Each TIMSS 2019 country and benchmarking participant prepared a chapter summarizing key aspects of mathematics and science education, and completed the TIMSS 2019 Curriculum Questionnaire."

It seems reasonable, therefore, to assume that the "key aspects of science" as identified by each country are represented in their descriptions of their science curricula.

Country	Ethics	Problem Solving	Critical Thinking	Scientific Thinking/ Scientific Method	Change in Scientific Understanding over time
Australia	✓	✓		✓	✓
Chile					✓
Finland		✓	✓	✓	
France	✓	✓	✓	✓	
Israel		✓	✓		
Italy					
Japan		✓		✓	
Kazakhstan					
Lebanon		✓		✓	✓
Morocco				✓	
Russian Federation		✓		✓	
Singapore	✓				
South Africa					✓
USA					

Mention of different topics in 14 countries who responded to the TIMSS 2019 survey

In all of the 14 countries I studied on the TIMSS site, only three – Finland, France, and Israel – mentioned critical thinking at all. A few mentioned problem solving. Six countries mentioned neither critical thinking nor problem solving. Seven countries mentioned neither scientific thinking nor scientific method.

Note that although Singapore mentions Ethics in a diagram, it does not appear elsewhere in the document. Similarly, I have given France and Russia check marks for Scientific Method though they don't actually name it, but they were the only countries that mentioned hypothesizing and testing.

If you exclude ethics, as apparently scientific curricula mostly do (and you have only to look at the state of worldwide politics to realise it's not only science education that has this problem), there are still no countries that cover all the topics explicitly. Even when we combine scientific thinking and scientific method, it doesn't get us very far. What's particularly disturbing is that only four countries make any reference at all to the fact that scientific understanding changes over time.

It is this aspect of the changing nature of scientific understanding that I suspect is key to many of the criticisms levelled at science in times of crisis, from climate change to pandemics. If you don't understand that science constantly tests and updates its understanding, then changing scientific "knowledge" or advice looks like failure. In an age where we are increasingly intolerant of politicians and public figures changing their minds on key issues, science apparently changing its mind is impossible to accept. To fix that, we need to teach science as a process. As a constantly updating way of understanding the world. Not as a matter of hard, unchanging facts.

I was schooled on twitter recently when looking for a quote, because I used the word theory incorrectly. I said that until a finding has been replicated, it's still just a theory. Professor Ray Wills pointed out to me that a *theory* is established science. A *hypothesis* is the yet to be tested idea. That's a crucial indicator of the nature of science, because in the real world we use 'theory' to mean an unconfirmed idea. Indeed, the Dictionary definition supplied by Google is that a theory is a 'supposition'. Not a fact.

And really, everything we think of as a fact is just a theory in science. A theory is the best explanation we can come up with for the observable data. And it doesn't graduate from hypothesis to theory unless it passes every test we can throw at it. There's always the chance that new evidence, from new types of instruments, new experiments, or new understanding of data, will *disprove* a theory. But you can't ever prove a theory. Only disprove it. So it never becomes a fact.

I don't remember ever being taught that. The first time I heard it discussed in formal education was when I was team teaching year 10 science in 2017 with an extraordinarily gifted and dedicated science teacher by the name of Kathryn Grainger. When I casually flung around the term "prove", she called me out on it. *"You can't prove a theory."* she remonstrated. *"You can only support it or disprove it."*

It was a wonderful moment, because our students saw critical thinking in action. They saw one expert challenge another expert. And they saw an expert with a PhD in Computer Science and more years teaching at both secondary and tertiary levels than any of them had been alive say *"of course! I was wrong! And that's a really important point!"*

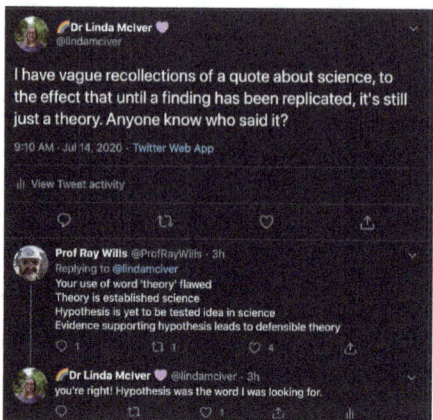

Viewing science as all about facts leads to a huge problem when the "facts" change. If the best scientific advice, on the evidence we have now, says one thing, and tomorrow we find more evidence that shows that yesterday's advice was wrong, people view this as proof that science is broken. But this is not broken science. This is actually how science progresses. Changing the "facts", or updating our understanding, shows that science is working.

This makes a lot more sense if you teach theories not facts, and if you base your science education on testing those theories.

Science in Schools: An Education in Confirmation Bias

Far back in the ancient mists of time, when my husband studied year 12 Biology (I did year 12 the same year, so I get to make those jokes), there was a question that caused some controversy:

True or False: all autotrophs contain chlorophyll.

According to the year 12 curriculum that year, the answer was True – every creature that creates its own energy uses chlorophyll to do it via the process known as photosynthesis. The trouble is, in the real world the answer is False. It turns out that there are teeny tiny beasties that live around volcanic vents in the deepest darkest recesses of the ocean, among other extreme environments, that use a process called chemosynthesis, using chemical reactions to fuel their food making.

This is one of the gotchas that teaching science as facts leads to, almost inevitably. Because the world is complicated, and teaching facts according to the theories you are up to in the curriculum is quite likely to fall foul of what we know according to more complex models that we haven't taught yet.

So we wind up assessing according to what's in the course, rather than according to how the world actually works. And it's worse than that, because when we teach facts and known processes, we encourage students to get the expected answers.

School science experiments involve known inputs, known processes, and known outputs, and are often assessed as reports that are required to contain the expected answers. This means that students are doing experiments where they know the results they expect to get, and they know they will be assessed according to whether they got those results or not.

Consider a chemistry experiment that involves taking a fixed amount of reagent A, adding it to reagent B, and measuring the time taken for the reaction to finish bubbling at different temperatures. Students get results and produce a graph, but it doesn't look right. What do they do?

In a world where science education was explorative and based on understanding, perhaps the students would be able to explore why they got different results. Maybe they would repeat the experiment with the same equipment, but also with different equipment, to see what happened. Maybe they would test the reagents, or recalibrate the thermometer to try to find out what really happened.

In practice there is very rarely time for this. Typically, they adjust the graph until it looks right, changing their data as necessary, often because the marks for the experiment hinge on the report that contains the expected results. We are teaching them to achieve a known outcome – in other words, we are teaching them confirmation bias: If you don't get the result you expected, change it until you do. This is the absolute opposite of fallibilism. It is not teaching our kids to test and be sceptical, and to explore what happened when they get unexpected results. It is teaching them to get the "right" result.

Lest you think this is a needlessly cynical view of science experiments, consider the story I told in the introduction, about a thermodynamics experiment that produced unexpected results. Temperatures that were expected to reach equilibrium actually crossed over a little. Those results could have been the basis of a wonderful conversation about testing theories, about data, and about exploring the unexpected. Instead, the curriculum effectively demanded that the results be sanitised so as not to "confuse the students". There were several teachers involved, and they all agreed that the best thing for the students was to delete the anomalous results.

Now what?

How do we change our education system to emphasise and measure the things we really care about, rather than focusing on the things that are easy to measure? To figure that out, we need to first understand the issues in education today, which leads us to Chapter4. Measurable or Meaningful: Pick one.

4

Measurable or Meaningful: Pick One

Picture a primary school assembly. Hundreds of kids file in on a Monday morning, in various stages of bouncing, wriggling excitement. They sit down on the floor as expected, but it's taking them a while to settle. The principal stands up with the microphone and sadly says "oh, children, that's not our Brandon Park Manners!" and the room is miraculously settled.

The school is Brandon Park Primary School, and this is as close as the principal, Sheryl Chard, gets to angry shouting. The kids adore her, and would do anything for her. She knows every name (as well as those of every parent and every sibling!), she takes an interest in everybody. Every child at that school is firmly aware that they are important, and that she expects them to be kind, to treat others with respect, and to work hard. And for her, they do. I'm not going to tell you that every child at that school is perfect, but the simple act of expecting the best goes a long way to ensuring that the best is what we see.

Sheryl's expectations may bring out the best in her students, but there are many ways our school system signals its expectations to our kids, and sometimes the signals are unintentional, yet extremely powerful. Few of those signals are stronger or more influential than assessment. Particularly final year assessment. In this chapter, I will look at how assessment shapes our education system.

A statement of priorities

In 2020, as covid19 sunk its teeth into the world, and governments scrambled to respond, the state of Victoria, in Australia's South, was relatively quick to move schooling online and head into lockdown, in an attempt to slow the spread of the virus before it got out of control. For the start of term 2, online schooling was the norm. Once the state's covid numbers were "under control", schools reopened and life went back to something approaching normal for the end of term 2, and we looked forward to a relatively normal term 3, starting in July.

Sadly, the covid numbers began to spike again, and the return to school never happened in term 3 at all, for metropolitan students. Melbourne went into strict, stage 4 lockdown. Only essential workers were permitted to leave home for work. Everyone else was banned from leaving a 5km radius from their homes, and even then only permitted to leave home to buy food, for medical care, or for one hour of exercise per day. Weirdly, though, there was one exception. Students in years 11 and 12, the final years of school in Victoria, together with any year 10 students doing a year 11 subject, were required to keep attending school, in order to avoid Melbourne students falling "out of step" with students who weren't in lockdown.

Unsurprisingly, school-centred outbreaks continued to occur, and on August 5th the decision was reluctantly made to send even year 11 and 12 students home to learn remotely[80], and close the schools completely.

As you can see in Figure 5, the graph of Victorian Covid Cases, the cases began to drop within a fortnight. Given that there were ongoing school-based infection clusters, closing the schools was absolutely necessary. The delay in closing the schools balanced public health (and indeed the health of the students and teachers themselves, not to mention their families and communities), against what seemed to be an urgent desire to maintain the validity of the final rank, the ATAR, to preserve the ability to fairly compare kids in metropolitan Melbourne with kids at school in country Victoria, and the rest of Australia.

The ATAR, or Australian Tertiary Admission Rank, is a value out of 99.95 that is used to judge which students are accepted into many university courses. The ATAR is calculated from students' scores in all of the year 12 subjects they study, but not the raw scores. Each student is given a "study score" for each subject, which is scaled according to various factors, including the school, and the relative difficulty of that subject.

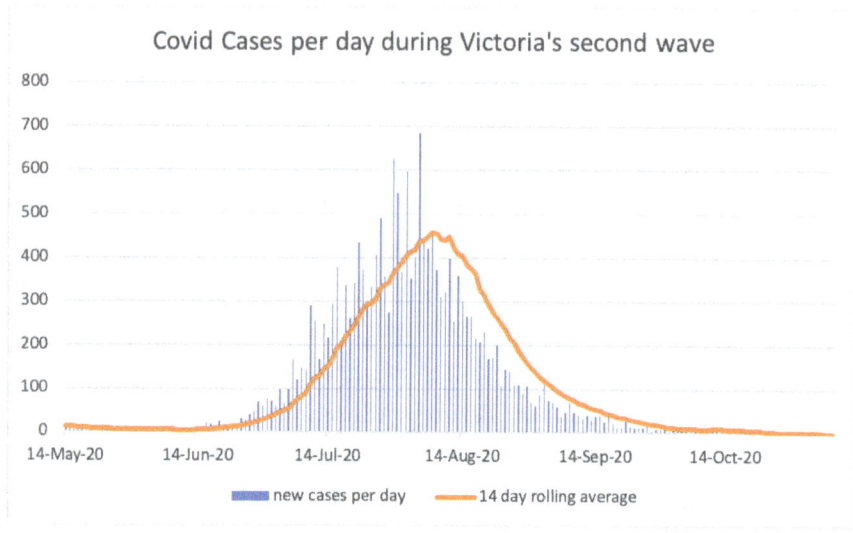

Covid Cases per day in Victoria from May to November 2020, the second wave

This is an interesting exercise in policy making and logic. In a deadly pandemic, which killed over 800 people in Victoria (and literally millions overseas), why was the need for comparability of the ATAR between metropolitan and rural students so urgent that it justified taking the risk of continuing to allow those students to travel to school – in some cases far out of their local areas – and mix? Particularly when every other Victorian was locked down and trapped in their homes for 23 hours every day?

It's worth noting that the ATAR is the primary – in many cases the only – means of determining entry into university courses. The higher the ATAR, the more choices of tertiary course a student has. The ATAR is a rank, and the highest possible rank is 99.95. That is, as far as I can determine, the sole use of the ATAR: To determine university entrance.

According to a study conducted by the University Admission Centre[81], ATAR is a reasonable predictor of a student's results in first year. The study acknowledges that, in some cases, ATAR does not reflect a student's potential, but concludes that it is an efficient and reasonable proxy, stating: *"However, for a large number of courses where the fundamental requirement of the student is to possess the right level of academic ability to meet the demands of the course, the ATAR is an effective tool in predicting the likelihood of this outcome."*

This seems to me to be a very telling sentence: It more or less declares that the academic ability to meet the demands of the course is the overwhelmingly important criterion. Yet, when I choose a doctor, her academic ability to meet the demands of the course are of almost zero interest to me. I am concerned about her communication skills, her ability to keep abreast of current developments in medical research and practice, and her willingness to work collaboratively with me as an active agent in my own care.

The problem with the ATAR is that it has become a goal in its own right, rather than a means to an end. As anthropologist Marilyn Strathern puts it, "When a measure becomes the target, it ceases to be a good measure." Which relates to the idea in Physics exemplified by the famous Schrodinger's cat experiment – that the act of observing something changes what you are observing. When we use standardised testing such as Australia's Naplan (or the ATAR) to measure a school, the school inevitably tries to maximise its score.

In 2017 I was lucky enough to hear Nobel prize winner Wolfgang Ketterle speak, and one thing he said absolutely nailed me to my chair. He told us that he made his Nobel Prize winning discovery when he was new to the field. Moreover, he felt that being new to the field was *why* he made the discovery, because he was seeing things in different, non-traditional ways. Yet our focus on exam results, on ATARs, on traditional, efficient measures like exam results and ATARS tends to mean that we are training our kids, as we were ourselves trained, to conform, respect authority, and be obedient. What if we were trained to be rational heretics instead?

What kind of people do we want our kids to be?

When I first sat down to write this book, I did a trawl of university websites around the world, to try to find out what they considered the most important attributes they claim their graduates will display.

Monash University in Melbourne, for example, says this[82]:

Monash University prepares its graduates to be:

- responsible and effective global citizens who:
    - engage in an internationalised world
    - exhibit cross-cultural competence
    - demonstrate ethical values
- critical and creative scholars who:
    - produce innovative solutions to problems
    - apply research skills to a range of challenges
    - communicate perceptively and effectively.

While Adelaide University has this list[83]:

- Attribute 1: Deep discipline knowledge and intellectual breadth
- Attribute 2: Creative and critical thinking, and problem solving
- Attribute 3: Teamwork and communication skills
- Attribute 4: Professionalism and leadership readiness
- Attribute 5: Intercultural and ethical competency
- Attribute 6: Australian Aboriginal and Torres Strait Islander cultural competency
- Attribute 7: Digital capabilities
- Attribute 8: Self-awareness and emotional intelligence

Most institutions publicise such a list, and they vary in the details, but the high level attributes are remarkably consistent. They say they want

- Creative
- Ethical
- Problem Solvers
- With Knowledge of a discipline

Which is interesting, because most university courses don't measure the first three of those points much, if at all. This is largely because assessment of knowledge of a discipline is easy – simply measure how many facts a student knows, how many of the procedures and processes they can apply. This is quite straightforward to assess with a standard assignment/exam combination.

Some academics, of course, do far better than the standard, and some courses both teach and assess ethics, creativity, and problem solving. Unfortunately they still seem to be the exception rather than the rule, and it's often a fight to get such courses accepted, especially in technical faculties. Once the staff member championing the course is moved on to teach other subjects, the subject is often quietly retired, or rewritten to be more "conventional", and easier for traditional academics to teach.

One of the key qualities we measure in assessment is validity – does the assessment actually measure what we think it measures? With facts and known processes & procedures, validity is relatively easy to achieve. There are always issues with exam validity. Some students don't perform well in exams, for example. If they freeze with anxiety under exam conditions, their answers to exam questions might not be very accurate measures of how much they know. Sometimes exam questions are ambiguous, or confusingly worded. Sometimes they don't ask what we thought they asked.

Reliable assessment is assessment that gets the same result every time. If you run the same assessment, with the same student, under the same conditions, you should get the same result. It's not the same as validity, but it's important. Most exams are pretty reliable. But do they measure what we want them to measure?

As Associate Professor Nick Falkner, from Adelaide University points out, getting 60% on a multiple choice exam does not mean a person knows 60% of the course, even in the rare event that the exam actually covers 100% of the course curriculum. Instead, results on a multiple choice exam actually show that when someone is presented with a limited set of options, their understanding and vocabulary have led them to pick this one. As Dr Falkner puts it, *"The real question is have they actually learnt something, or have they been trained in pattern recognition in a limited space?"*

Academics often ruefully say that "Knowledge is not transferable between semesters," which raises another concern with this form of measurement – do the students actually remember the facts for any significant length of time after the exam?

To summarise, we tend to use exams because we have been using them for a long time, because they are relatively easy to write and to mark, and because they are fairly reliable. And, of course, because they allow us to rank students in what we can pretend is an objective way.

But do they measure creativity?

Possibly the big problem with assessment "instruments" such as tests and exams as measures of learning is that we assume learning can be tested in a mechanised, human-free fashion, whereas true assessment of where a student is at requires interaction with the student. Of course, this introduces the risk of bias. But the notion that tests and exams are bias free – even if they are blind marked–is somewhat farcical anyway. Students who interpret things differently can be penalised by exam questions that are rote marked. For example, my daughter, Zoe, once had this maths question on an assignment: You have $300 in your wallet, which you deposit into your bank account. Is this a positive or negative transaction?

Most of the kids said it was negative, because the amount in your wallet goes down. But Zoe pointed out that the question was ambiguous: It's a negative transaction from the perspective of the wallet, but a positive one from the perspective of the bank account, and a neutral one from the perspective of yourself–you have the same amount of money you had before, you're just storing it differently. When she took this problem to her maths teacher he sympathised, said she was right, but told her to put the "right" answer (according to the answers in the teacher version of the text). Had Zoe had that question on an exam, she would very probably either have given the wrong answer, or wasted valuable time trying to make sense of it, when she could have been answering other questions. That question was effectively teaching my daughter to pass assessments, not to think critically, logically, and carefully.

The problem of making sure you teach and assess the things you want students to learn is something of an elephant in the room of the teaching profession. When you are under extreme time pressure, as most teachers are, it is natural to reach for tried and tested techniques – to do things the way they were done when you were at school. And to teach things the way you were taught to teach them. Where evaluation of teachers exists, it tends to reinforce this approach.

Years ago I got into an argument with Ted[84], the leader of the faculty I was working in at the time. I argued that the course we were teaching was fundamentally flawed. The students we were teaching were not learning what we wanted them to learn, and they couldn't see the point of the subject. Ted argued that the course was great, the kids loved it, and we didn't need to change a thing. I wanted to run an anonymous survey to find out for sure, which he finally agreed to – with one small flaw. He didn't make it anonymous. Out of 200 students he got 20 replies, and, what do you know? They said "the course is great, it doesn't need changing".

Ted was outraged when I suggested that collecting the kids' identities made it unsafe for them to say what they truly believed, directly to the teacher responsible for the course. We argued a lot, and in the end we asked a neutral party – a teacher from another faculty – to run a focus group. Now the kids were reporting on the subject to someone who was uninvolved in it, and uninvested in the outcomes. And what they said was horrifying to Ted, but no surprise to me. They hated it. They couldn't see the point. Some of the kids in the focus group had also answered the survey, but their responses were very different.

That survey did not answer our questions. What we wanted to know was "what do kids really think about this course?" and what we wound up asking was "what will kids say about this course when asked by the guy who designed it, and when he knows who they are?"

Had we based the continuing shape of the course on that survey feedback, we would have been basing our ideas on data that didn't say what we thought it did. This is one of the big problems with all data, not just assessment data. No dataset is perfect. No collection technique is foolproof. It's very easy to ask leading questions on a survey that get you the result you are hoping for (such as: "how awesome was the course?" Rather than "how did you feel about the course?"), to survey a subset of people that don't represent the entire population, or to assume the data is complete when there are significant parts missing. So we need to be super cautious when we use data to justify our actions, and to shape changes in our systems. Does the data say what we think it does? How can we be rationally sceptical and test our assumptions?

Our education system is heavily measured. Between standardised testing such as Naplan and PISA, external final year exams, and all of the performance measures imposed on teachers, we are measuring outcomes constantly. Unfortunately, we don't seem to pay a lot of attention to the question of whether those outcomes are the ones we really want to aim for.

We're looking for assessment reliability – will we get the same result if we do the same test again? – rather than validity – are we measuring what we think we're measuring, or, indeed, asking the question: what we *should* be measuring? And that, in a nutshell, is the issue with education. We shape our education systems to maximise outcomes. Unfortunately, the outcomes we are maximising are PISA[85] scores and exam results. In an ideal world, these would be measures of learning but, as we will see, they don't always measure what we think they do. The other issue, of course, is that we also shape our education systems to minimise cost.

The myth of perfect data

Any dataset we work with has flaws. Usually, they are not exactly the information we want, rather they are as close as we can easily/cheaply/quickly get to that data, or simply the data that we have access to. For example, population datasets are nearly always data from a sample of the population, rather than from everyone, which means some people will not be represented by that data.

Consider the census data for Australia, which covers the population who filled out the census. This will not necessarily cover people who were homeless at the time (because census forms are delivered to fixed addresses), or overseas, or who chose not to fill out the form. These people might leave significant gaps in the data if it is being used to calculate, for example, how many hospital beds we might need in a particular area.

The census data is the data we have. What we want is how many people live in an area, and how that might change over the next four years.

What we have is how many people filled out the census on census night (and the previous census nights). This is, ideally, a close approximation, but it's not the same. In 2016 it was a very poor approximation indeed, because the census was online for the first time in Australian history, and the system actually crashed under the load on census night[86]. There were also widespread concerns about the privacy and security of the system, and the information collected. Not to mention barriers to access for people with no internet connection, or for whom English is not their first language. Some people chose not to fill out the survey at all, or to use false data to protect themselves[87].

Some of the data we want is just not easy to measure. For example, we want to measure kids' learning, so we have them sit exams. They are great for measuring recall of facts, and application of known procedures. They are rarely used to measure problem solving, creativity, or ethics – the attributes we say we care about. Plus they don't necessarily even measure recall very well, depending on how well the exam was written, what the conditions were on the day, how good students are at doing exams, etc. We tend to use exam results as a proxy measure for learning, especially when we use those results, for example, to decide who gets into a particular university course.

Unfortunately in education we sometimes forget that what we are measuring is not actually what we want to know. We tend to shape our education to be measurable, rather than to be meaningful.

All datasets have issues like these. The challenge is to identify the issues, and take them into account when we're using the data to shape our future.

The easiest way to assess students in schools is to use things like multiple choice questions, which have been computer markable for decades – this is why students have to fill in some kinds of tests with very particular types of pencils, or particular coloured pens, and by colouring in circles on a page rather than writing their answers. It makes the tests simple to feed in to an automated marking system. Now, of course, we can do that online even more easily.

Multiple choice questions are very simple ways to measure students' knowledge of facts and known processes. You can even ask reasoning style questions, though they are more difficult to write in ways that all students can understand and interpret correctly, and, unlike a written answer, multiple choice gives the teacher no room to spot any misunderstanding that they didn't see coming.

Doing this kind of assessment produces a very convenient number that you can effectively stamp the student with. "You received 90% on this test, therefore you know 90% of the material taught to you this semester." Of course, that relies on the test testing 100% of the material taught, and on every question being correctly written, and no tests using questions like the photosynthesis question mentioned in chapter 3, that have one answer according to the course material taught so far, but another answer (or multiple possible answers) in the real world. And, for the most part, these tests are a test of a student's memorisation rather than their understanding.

I've lost track of the number of teachers I've heard talk about their teaching goals and say things like: In the literacy program, our goal is to improve the Naplan and VCE results. This is heartbreaking, because the goal of a literacy program is surely to improve literacy, not to improve the Naplan score. Naplan literacy results are intended to be a measure of literacy, but they are actually an accurate measure of how well students do on the literacy section of the Naplan test. Ideally this will be close to a reasonable measure of literacy, but we always need to remember that the measurement is not actually the same as the thing being measured.

You might think, then, that the important question is this: How do we shift assessment so that we are assessing the things we really care about? Things like ethics, creativity, problem solving, and logical reasoning. And this is an important question. But I actually think we need to take the argument one step further, though it feels extremely heretical[88]: What is assessment *for*?

What is assessment for?

What even is assessment? If you ask a bunch of teachers to explain assessment to you, you will likely get an explanation of summative versus formative assessment. The short version of that is that summative assessment gives you a measure of how much the student knows or can do – a *sum* of their learning – and formative assessment is a kind of feedback to the student to help them improve – it helps *form* their learning. Summative assessment often forms the end-of-course grade, while formative assessment happens throughout the course, to highlight gaps in a student's knowledge, and to identify "skills they are great at" vs "skills they need to improve in". Sitting a practice exam counts as formative if the student gets the opportunity to go through the results and note the types of questions they got wrong. Sitting the final exam is purely summative, unless the student has the opportunity to go through the marked exam afterwards and learn from their mistakes, in which case it can also be formative.

Formative assessment is obviously useful to students trying to improve their performance, but the question is whether they are trying to improve those important attributes of creativity, problem solving, and ethical behaviour, or whether the formative assessment is more along the lines of practice exams: trying to improve their performance on the all-important summative assessment that determines which university they can go to, and what courses they can study. When the focus is on summative assessment, students often become obsessed with a single mark lost, or with whether they did better than the other kids in the class.

Summative assessment is often used to judge students, to rank them relative to their peers, and to determine entry into subsequent courses or degrees. The ranking aspect can be particularly problematic if there are factors that vary between, or indeed within, cohorts. Consider the ATAR, or Australian Tertiary Admission Rank. It is clear from the data that there is benefit from being at particular schools, sometimes even in particular classes[89]. From not getting sick or not having traumatic family circumstances. From being in a metropolitan school rather than a rural one. There are many factors that the ATAR endeavours to compress down into a ranking that effectively says: This student is more likely to do well in this degree than a student with a lower rank. It's a way of saying "*This* student makes it in. *That* student doesn't." and being able to justify it with a nice, "objective" number.

The trouble is, the number might not be as objective as all that. Many studies have been conducted to try to determine whether Socio Economic Status (SES) has an impact on ATAR. Like a lot of educational research, it's difficult to find a definitive answer, because you can't control conditions entirely. However, a recent study by Emmaline Bexley[90] and her colleagues reports that high Socio Economic Status (SES) students who were achieving similar grades to low SES students in Naplan in Year 9 went on to achieve ATARs around 10 points higher than the low SES students three years later. Given that the maximum ATAR is 99.95, that's over 10% of your score that's closely related to your socioeconomic status. It may not be causative – there are many factors impacting low SES students that might impact their ATAR, such as the need to work part time to supplement family income, potentially insecure housing, etc – but the fact that the correlation exists is, in itself, disturbing. If we have any commitment at all to equitable educational outcomes, this has to change.

Policy makers who are in favour of a particular approach, such as standardised testing, have a bad habit of using their own confirmation bias to support their preferred approach. In other words, they look for evidence that tells them what they want to hear, and ignore evidence to the contrary. As Finnish Education researcher Pahsi Sahlberg points out: *"Evidence-based education policies use research to link selected treatment and expected outcomes, but they almost always ignore possible harmful side effects they may have on schools, teachers or children. Take NAPLAN, for example. Those who advocate the necessity of national standardised testing regimes back their views by positive consequences of high-stakes testing while ignoring the associated risks that research has exposed: narrowing curriculum, teaching to the tests, and declining student motivation, just to mention some."* [91]

The end product of the Australian school system for most kids is the ATAR, and which university course they can get into with it. A particularly disturbing aspect of this focus on a final ranking is that kids often choose – and are encouraged to choose – subjects in which they are more likely to do well, rather than subjects that they are actually interested in.

The other obvious problem with this system is that what we are doing is training kids to be very good at exams. We then use how good they are at exams to select the courses they will do, and those courses mostly use exams to determine how well they do in those courses. So we are selecting kids to train to be engineers, doctors, architects, lawyers, teachers, scientists, etc, on the basis of how well they do in exams. And then we rank them as engineers, doctors, architects, lawyers, teachers, scientists, etc on the basis of how well they do in exams. And then we send them out to be engineers, doctors, architects, lawyers, teachers, scientists, etc, where they will be required to do a huge range of things that bear no resemblance to sitting exams at all.

If we truly want creative, ethical, rational, critically thinking problem solvers, then it makes sense to ask if our school system is actually producing kids with those characteristics. It's not clear that we're even turning out kids who *value* these kinds of characteristics. The system currently runs on marking criteria and constrained outcomes that punish the kind of kids who see a problem with the assignment definition and create a whole system to solve that problem. The kids who misinterpret exam questions because they think laterally. The kids who know that there are creatures in extreme environments who produce their own energy without photosynthesis. The kids who solve problems differently and come up with creative solutions, but that don't fit the rubric. The kids who write more, like my year 11 student Chris, who struggled so much with the word limit on an assignment that he made a separate web page to add in the in-depth data that explored the topic in so much more detail. We teach kids not to do that. We teach them to meet the criteria and stop. Do the minimum.

Chris was also one of the students who took his year 11 Computational Science project, together with his partner Matt, and continued working on it in year 12. Chris and Matt were producing software to enable a cancer researcher to do more powerful and effective research. But they received no credit for it. There was no room in the ATAR for it. Indeed, their teachers advised against putting too much effort into it (despite it being used by a scientist to contribute to cancer research), in case it distracted them from the really important thing – not cancer research, but receiving the highest possible ATAR. What kind of values are we teaching with that kind of message?

What do we WANT the purpose of education to be?

So, here we are. Assessing our students mostly using exams and assignments that have no purpose other than to contribute to assessment (and maybe, if we're lucky, to learning). From the early school years right through to undergraduate courses at university, we are teaching kids that what matters above all is the final mark. Ethics, critical thinking, community service, even health and wellbeing, and yes, even learning, become subservient to that final mark.

Consider this google search result, when I searched for "atar future". "Enter your ATAR to browse the courses available in your range." The clear message here is: Do not look at courses for which you do not have the ATAR. They are not for the likes of you. Don't go considering alternative pathways, or how you might study something related, do well, and transfer into your course of choice. Be realistic. Accept your limitations.

When I was teaching, one of my students told her form group teacher that she was aiming to get into one of the big tech universities in the United States. CMU, Caltech, MIT, or somewhere like that. He talked her down, told her she needed to be realistic. She's now studying at MIT. Another student wanted to do a subject in year 12 for which you needed really strong marks in year 11. He wasn't on track to get those marks, so his form group teacher also told him he needed to be realistic. That it wasn't something he could do. He came to me, distressed, and I took a slightly different line. I said he could do it, but that it would be tough. He'd have to work super hard. And there were no guarantees. He did not, in the end, get the marks he needed to do that subject. But he had worked harder, reached higher, and developed some confidence from the fact that I believed in him. (I still do.)

www.deakin.edu.au › choose ▼
Find your future | Deakin University
Discover Courses by **ATAR**. Enter your **ATAR** to browse the courses available in your range.
Filter: **ATAR**.

Ad for Deakin University: Discover courses by ATAR

I actually think that judging his ability to do that subject by his marks was foolish. It was an attempt at objectivity. If we make decisions based on a number, then surely we can't be accused of bias, or prejudice? It is a way of wriggling out of the ethical and emotional complexity of a decision. How do we choose which kids are given the chance to become doctors? How do we choose which kids could become engineers? How do we select teachers, nurses, or data scientists? We remove ourselves from any possibility of making it personal – no-one can say "She just didn't like me", or "He didn't like the colour of my skin", or "She didn't let me in because I'm a girl." It's all down to this simple number. Objectivity guaranteed.

But if there is a correlation between socioeconomic status... if girls are driven out of particular subjects by the perception that they are not suited to them... if rural kids don't have access to the same range of subjects... if some schools don't have great teachers or support structures... then what we have is the pretence of objectivity and fairness, rather than actual objectivity and fairness.

As we saw with the Biometric Mirror in chapter 1, numbers and computer outputs provide a veneer of respectability that in many cases is wholly unwarranted. You could call it the cloak of objectivity. Once you throw on the cloak, no-one can argue with your actions. Add in university courses with ever-increasing fees – my daughter is starting first year university as I write and has a "Commonwealth Supported Place", meaning her course "only" costs over $8000 for her first year alone – and you find that the university courses available to you may well be as much determined by your postcode and bank account as they are by your actual abilities.

I don't want a doctor who got into the course because they were rich and went to a high socioeconomic status school. I want the best doctor I can get. Or Engineer. Or Teacher. Or Lawyer. Or Accountant. Or Data Scientist. If we exclude – or allow the system to exclude – people from courses due to their background and economic status, then we risk losing some of the best people for the job.

Interestingly, many degrees and institutions are coming to realise that the ATAR does not give them the information they need to determine the best candidates. There is a marked shift to using interviews in addition to the ATAR for many courses, including medicine.

One study at Queensland University found that removing the interview from the selection process actually increased the gender bias – taking men from 50.9% of students to 64%[92]. Men, apparently, consistently perform better on section III of the GAMSAT, the entrance exam for medical degrees. While it's not proof, there is a strong implication that section III of the GAMSAT contains some form of gender bias. It also suggests that our understanding of the assessment process remains flawed, if we can't reliably write an unbiased, equitable exam.

Assessing the Assessors

When I was an academic, every course that was taught had to go through a student evaluation process. There was a lot of numeric data collected in the process – rankings of different criteria like the quality of the teaching, the quality of handouts and lecture notes, etc, but I always found that the written comments were the most illuminating. Many lecturers referred dismissively to these evaluations as popularity contests, and, it turns out, they weren't wrong. In recent years, multiple studies have been conducted that show that these evaluations are subject to significant bias against women and people of colour.

There are clearly complexities in evaluating teaching quality. But evaluating a new course seems like basic science to me: if you do something new or change something, compare it with what you were doing before, or if there's nothing to compare it with because it's completely new, at least do a baseline evaluation to figure out what worked and what didn't, so that you can compare the next version with this first one.

It always amazed me that, as a new lecturer, there was no-one checking over my shoulder to ensure that I was doing a good job, or even teaching the right materials. I was incredibly fortunate to have an extraordinary teacher, Damian Conway, as my PhD supervisor, so I had picked up a lot of really good teaching habits, but there was no formal process of teaching evaluation, monitoring, or even training. These days many institutions have some kind of formal qualification in tertiary education, but there is still very little oversight of actual teaching practice. There are, at least, standard teaching evaluation surveys that run at the end of each semester. It's not ideal, but better than nothing.

When I made the switch to high school teaching I was further amazed. Once again, there was no quality control. No-one looking at what I was teaching. No-one checking to see if my teaching skills were up to scratch. No-one overseeing what went on in my classroom at all. I was lucky that I started teaching in a school with team teaching, which meant that I could watch what did, and didn't, work with my teaching partners, but in classes that were singly taught, there was no oversight at all. Every year we'd do a "performance development process" that was not so much about evaluating and improving our performance as it was about writing legalistic documents to make it look as though we were evaluating and improving our performance. Bad teachers – there are always some – had no incentive or support to improve.

Workloads

Lest you think I taught at a particularly bad school, let me assure you that this is quite normal. There is very little active performance management in schools. To be honest, there is no time for it. A few years ago when I was still teaching, after someone told me that I had a cushy job, I did some calculations about how I was spending my time. This is an excerpt from the blog post I wrote about it:

I am paid for a 19 hour week (half of the standard 38 hour week). That's 1140 minutes. Of that time, I teach scheduled classes for 675 minutes. We have 75 minute classes, so I frequently teach for 150 minutes (with a five minute transition time that just allows me time to get to the next room), then get a 50 minute lunch break, followed by another 75 minute class. (Bear in mind that I can't leave to get a cup of tea or even go to the loo in class time, as I am on duty and required to maintain minimum staffing ratios in that room.)

For my 2.5 work days, I get 50 minutes lunch break a day – 150 minutes in total. This is "my" time, so on Tuesdays I help with the choir, Thursdays and Fridays I meet with students who need extra help, as well as doing a 25 minute yard duty. If you add those "free" times, together with the 25 minute tea breaks in the morning, also usually spent on yard duty or helping students, we're up to 900 minutes. After school on Tuesdays I meet with my teaching team for up to an hour, planning curriculum, organizing competitions, planning excursions, and making sure we are all teaching the same things. Now we're up to 960.

On Wednesdays we have professional learning in the afternoon, but I'm only there for 50 minutes of that once a fortnight, due to the way my hours have worked out this year, so let's call it 25 per week. 985. Thursday afternoons I run an hour of extra programming help for my year 11 students, where they can ask questions, get help with particular problems they have, and go over some of the trickier stuff that they might not have fully understood in class. 1045. Not including those extra meetings that arise when excursions need to be organized, or extra activities run, like competitions, guest speakers, training sports teams, organizing school events etc. It also doesn't include attendance at Parent teacher interviews (after hours), school formals, open nights, presentation night, valedictory dinner, etc. All of these events come out of my own personal family time. Oh, and school camps, which we are expected to attend, but of course there is no such thing as time in lieu for non-work hours spent at work.

So that leaves me with 95 minutes of my working hours, per week. 95 minutes to plan 7 classes (2 of my face to face classes are covering for teachers who are away, so somebody else plans those), mark assignments for 78 students, track the progress of 78 students. Contact the parents of any students who are struggling. Meet with those parents to try to plan a way forward. Meet with students who have particular issues. Catch up with students who are no longer in my classes but will always be my students, who come to me for advice. Keep up to date with advances in my field. Plan new classroom activities and learn about new ways to engage my students. Meeting the teachers I team teach with to make sure we are on the same page for upcoming classes. Writing progress reports and end of semester reports. Completing mandatory Education Department requirements, and doing enough professional learning to maintain my registration. And a hundred other activities I haven't even got time to remember, much less complete.

Let's cut that to the bare minimum, throw away all those extraneous activities, and assume that the 95 minutes is half marking, half planning. And we'll round up, to be generous, and say 48 minutes for class planning. That's 7 minutes planning per 75 minute class. As to marking, I have 78 students on my rolls. That's around 37 seconds per assignment, assuming no toilet breaks or time to breathe. To be fair, that assumes that every student submits an assignment every week, which of course they don't. But they all do work every week, which I need to check on to ensure that they are making progress. And those students who don't submit their work need to be followed up on, to find out why, and put special measures in place to ensure that the work does come in eventually.

Think of the children in those classes. They very obviously deserve classes that have been designed in more than 7 minutes' planning time. They deserve detailed feedback on their work. They deserve plenty of time for one to one support outside class. And, for the most part, they get it, because teachers work insane hours to make it happen, leading to a crazily high levels of burnout, mental health issues, and teachers leaving the profession.

You could argue that teachers should be using some of their 12 weeks holiday per year to do planning, and most teachers certainly do, but you can't completely plan a course in the absence of the students who will study it, if you want to have a really effective learning experience. You need to be able to adapt the course to how the students are coping, what they are most interested in, and the things they are struggling with. And, of course, nobody who cares about their teaching works to rule and downs tools at the 38 hour mark. Most professions work more hours than they are paid for, and teaching is on the extreme end of that – which flies in the face of studies that show that working longer hours actually decreases productivity and increases error rates.

I once heard teaching referred to as FIFO work – Fly In Fly Out. In other words, during term time there is no time for anything but teaching. During the holidays, there is no energy for anything but collapse.

The practical upshot of all of this is that we give teachers workloads that are unsustainable, deprive them of the time and resources they need to do their best work, and then complain that results on standardized tests such as the Programme for International Student Assessment (PISA) test are declining.

Out of field teaching

As we saw in chapter 2, another thing that shocked me when I started teaching is the assumption in Australia that teachers can teach any subject. I taught in a very well resourced public school, and we had a Computer Science subject being taught by an English teacher, a Music teacher, a French teacher, and some Maths teachers. I've seen teachers who don't speak Mandarin "teaching" Mandarin at a primary school. I've seen Maths-phobic teachers, with no maths background since their compulsory maths at school, forced to teach Maths.

I assumed this was a rare and regrettable circumstance, but it turns out it's very common. It seems to be an inevitable consequence of the desperate funding shortage most public schools experience. Schools have to match the number of teaching staff to the classes to be taught really closely, because having any extra teachers would decimate their budget. Teachers are generally qualified to teach one or two subjects. So the chances of the mix of available teacher skills precisely matching the classes that need to be covered are pretty much zero. And that's before you add in timetabling complexity!

Think of it like a sliding numbers puzzle. With a gap it's difficult, but not impossible to move the number tiles into their correct places. Without a gap it's impossible. The gap, in this case, would be spare teaching capacity – meaning not every teacher has a full load, and extra teachers with particular expertise can be hired to fill in the gaps.

There are multiple ways the gaps could be filled. If there was a Universal Basic Income and working full time was not essential in order for most people to earn enough to afford to live, teachers could be flexibly part time and work enough hours to cover the classes for which they were actually needed, rather than having to work full time. This would also make it easier for teachers to spend the time they need for marking, preparation, and admin without sacrificing family time and a social life in order to do it. (Many teacher friends of mine simply do not socialise in term time, because they lack both the time and the energy to do so.)

Measurable AND meaningful – how can we have both?

The bottom line is that if we want to teach the things we say we care about – creativity, ethics, problem solving, and collaboration – then we have to show the students that we do, indeed, care about them. We have to stop using an ATAR based largely on assessments that don't, and probably can't, measure those things.

The good news is that we can absolutely do that, by giving students the opportunity to learn by solving real problems. By teaching them to critically evaluate their own results, and by optimising for carefully tested outcomes, rather than right answers, we can build habits of critical thinking and scepticism that will last a lifetime.

In Chapters 6 and 7 we will look at examples of projects like these, but first we have a hurdle to overcome. These projects don't have right answers. They can't easily be measured by exams or captured in a precise rubric. Which means we have to get comfortable with uncertainty. With not knowing where a project will take us. With not knowing what a great solution looks like. And with accepting that there is no such thing as a perfect outcome. That means that both teachers and students have to get comfortable with uncertainty. Chapter 5 looks at why uncertainty is important, and how we can come to terms with it: in education, in science, and in everyday life.

5

Accepting the Unexpected

The first time I gave lectures at university I was filling in for my supervisor, Damian Conway, and teaching some second year lectures on the C++ programming language. I wasn't confident with C++, and I was very nervous about the lectures, but Damian's slides were excellent, so I figured it was something I could do. There was one student who was clearly sceptical about my expertise, and he gave me a pretty tough time, right up until he asked a question I didn't know the answer to. I was so desperate not to lose face that I made up an answer, so that I could pretend I knew more than I actually did. I went home and researched the answer and discovered it was exactly the opposite of what I had said. The next day I went and stood at the front of that lecture theatre and confessed my mistake. I made the right answer clear, explained why it was the case, and went on with the next lecture in something of a funk, convinced that I had lost the class's respect irrevocably.

To my intense surprise, something very weird happened. The student who had been giving me a hard time started treating me with respect. He even came and thanked me for correcting that answer. The rest of the classes went much more smoothly, and the whole group seemed to relate better to me now. It was a revelation that shaped all of my teaching in the future. You might think that the lesson was "never bluff when you don't know the answer", and that was certainly one key message, but the really important thing that I took away was that it's ok to make mistakes, and you don't have to know everything. Indeed, you can't. No-one can. People actually respect you more when you admit that you don't know something.

Unfortunately, the model of teaching that I grew up with – indeed, that most of us do – is that the teacher is a dispenser of facts and wisdom. Teachers tell us stuff, we learn it. Which is great if you have a fixed course of facts that you want kids to memorise, but not so effective if you have problems that need solving, boxes to think outside of, or new ways of doing things to discover.

When I was teaching in a school, my first lesson with a new class I'd always stand up the front and, after all of the introductory bits were over, ask, "Do you trust me?" My students, who were mostly kind and well behaved people, would enthusiastically say "Yes, Dr McIver!" and then practically fall off their chairs when I'd say "well stop it!"

Of course, I did want their trust. But what I was looking to shake up was their blind faith. I wanted them to challenge me. To question what I told them. To look for errors, complexity, and deeper meaning. To ask difficult questions, raise challenging issues, and contest the status quo. And this was in a Computer Science class, where the binary of right and wrong would seem a logical fit for the content.

That dogmatic, black and white approach, though, is both dangerous and unproductive. It stifles creativity and makes problem solving a matter of rote pattern matching, rather than true innovation. Most of my lessons were discussion based, and the best lessons were often when students told me I was wrong. Everyone being in furious agreement doesn't make for good learning. People contesting accepted truths is vastly more interesting and educational for everyone.

It's confronting for many of us to not merely accept that we might be wrong, but actually invite people to figure out where we've gone astray. To stand up and say "I'm not perfect. I'm going to make mistakes. I'm going to be wrong. Look out for it! If you think you've found one of those times, we all *need* you to speak up."

All of these challenges are even more relevant to projects and assignments. If you've been raised in the traditional model of teaching, where the teacher is expected to have godlike powers of omniscience, and to never show weakness, it's hugely challenging to undertake projects that have no right answers. But the problems we need to solve in the real world do not have textbook solutions and predefined right answers, and we won't always know if we've solved them in the best possible way – or even in a way that works! So we need to figure out how we can make the shift from facts and omniscience to exploration and problem solving – and most importantly, to critically evaluating our own work.

Knowing versus Understanding

In Chapter 4 we defined science as the search for understanding, and made the bold claim that you cannot prove a theory. You can find evidence that supports it, or you can disprove it, but you cannot prove it. Everything is contestable. Everything we know could be wrong. Science's job, then, is not to look for proof, but to rigorously test every theory, to do its damndest to prove it wrong.

To be comfortable with this view of science, we must come to terms with uncertainty, and that is a fundamentally uncomfortable idea.

There is a lot of comfort in knowing things. It's so simple to be able to say that the sky is blue, that drinking a lot of water is good, or that fat, salt, or carbohydrates are bad for you. It's irritatingly messy to have to explain that, actually, the sky just appears blue for reasons that turn out to be remarkably complicated[93], or that drinking too much water can actually kill you[94], that we need a certain amount of fat, salt, and carbohydrates in our diet, or that "good" cholesterol and "bad" cholesterol are both essential for life.

Simple answers and certainty are very attractive. It feels like they give us a strong foundation to build on, and they make life feel easier. Unfortunately, the thirst for certainty can be a trap. For example, in the 1830s a mathematician by the rather impressive name of Lambert Adolphe Jacques Quetelet wanted some measure of whether a person could be considered obese or not. He devised a thing he called the Body Mass Index[95], which has been used ever since as a convenient way to categorise people as underweight, healthy weight, overweight, or obese.

Numbers, as we discussed in earlier chapters, give us a nice sense of objective certainty. They're not subjective the way simply looking at someone might be, or at least they don't *seem* subjective. The BMI is calculated by dividing someone's weight in kilos by the square of their height in meters. There's no room for subjectivity there at all, no room for nuance or complexity. Height is easily measured, so is weight. What could possibly go wrong?

The accepted categories typically follow the following template: 18.49 or below is underweight. 18.5 to 24.99 is a healthy weight, 25 to 29.99 is overweight, and 30 or more is obese[96]. The trouble is that those categories are not based on science. They're not based on any form of evidence at all. In fact, in 1998 the United States National Institutes of Health arbitrarily changed the definition of obesity[97], so that many people "became obese" overnight, (or rather, were redefined as obese) without a shred of evidence that they were actually unhealthy. The measure itself is simply a calculation devised by a statistician, with no basis in health, medicine, or science. There is no reason to give it any "weight" at all.

The BMI is a poor measure of health and fitness for white men, the people largely used for the creation of the metric, but it is even worse for women, people of colour, Asian people, and many, many others, because they weren't represented in the original dataset at all.

Unfortunately, the beguiling nature of this simple metric – human beings are, as noted in previous chapters, terribly susceptible to the appeal of reducing other human beings to a single number – means that despite evidence of significant flaws[98], the BMI is still used by doctors, hospitals, and government health organisations. Even the World Health Organisation treats it as a valid, meaningful metric[99].

A large part of the attraction of the BMI as a metric is the human craving for simplicity and certainty. Given the way we teach, and especially the way we teach science, it's no surprise that the idea that simple facts are all we need remains widely accepted. As soon as we pin a number on something, we give it a validity which can lead to it becoming embedded in our thinking, even though it might be wholly unwarranted.

And, of course, certainty, or at least trust, is important. As I write this, in April 2021, Australia is in the middle of a chaotic start to its covid19 vaccine rollout, and panic is erupting over possible blood clotting side effects of the Astra Zeneca vaccine, which the vast majority of Australians were scheduled to receive. People need simple answers to the question: Is it safe? Without a loud, clear, YES, many people will not be willing to risk being vaccinated.

Is it reasonable to expect the general public to be able to reason their way through the nuanced complexity of vaccine safety? There is a relatively simple calculation here, which is that the risk of death, severe illness, and possible long term complications from covid19 is much higher than the risk of complications from the vaccine. But though the calculation is simple, the components are not. Our understanding of the risk of complications from the vaccine is changing on an almost daily basis.

Unfortunately, to wrap our minds around that, we must first accept that very few things are entirely safe. It is safer to get the Astra Zeneca vaccine than to drive anywhere in your car. In 2018 Australia had a road death rate of 4.54 in 100,000 of population. That means that for every 100,000 Australians, 4.54 people died in that year.

By contrast, for every 1,000,000 people who receive the Astra Zeneca vaccine the European Medicines Agency (as of April 23rd 2021) estimates that one person will suffer the blood clotting response, which initially had a 25% death rate (note that now the condition is both detectable and treatable meaning very few deaths even among those who have the reaction), meaning a total death rate for the vaccine of 0.25 in every 100,000, or 1 in every 400,000. Yet we accept the risk of death from driving – 16 times higher than the risk from the vaccine – as reasonable, while the risk of death from the vaccine is terrifying.

It would be unethical for a scientist to state that the vaccine is completely safe, but it would be just as unethical to state that driving is safe. Yet the public, understandably, wants certainty. The lack of certainty is compounded by the fact that the statistics change as more people receive the vaccine. There will likely be other side effects we don't know about yet. Like most real world scenarios, the situation changes rapidly, and we have to be able to adapt and change our plans and behaviours as more data comes in. Right now I am keen to get the vaccine as soon as I am allowed, but who knows what will be different by the time I am eligible?

Would this conversation be any different if we were more comfortable with uncertainty? If we had grown up with projects with unpredictable outcomes, and problems without textbook solutions? Unfortunately I don't have a world in a petri dish where I have changed our education system and nothing else so that I could see what happened. (How useful would that be, seriously???) There is a whole field of research that deals with the psychology of risk perception, and our understanding of statistics plays only a small part in our responses. But if we were more used to the way science actually works, and the way new data becoming available changes our understanding, perhaps we would have better ways of understanding the changing information about the risks and benefits of vaccines rationally.

Getting comfortable with uncertainty

Getting comfortable with uncertainty has several aspects. One crucial aspect is qualitative data. It is easy to dismiss qualitative data as subjective and unimportant, but it is actually crucial to understanding much of our world. Sometimes we try to convert qualitative data to quantitative, by, for example, asking people to rate their pain on a scale of 0 (no pain) – 10 (the worst pain imaginable), or asking them to rank how the usefulness of a workshop they just experienced, from 1 (completely useless) to 5 (extraordinarily useful). These give you numbers you can work with mathematically, and compare from one workshop to the next, or one Physiotherapy appointment to the next. The trouble is that we tend to forget, having given something a score, that the measurement is not actually the thing being measured. A pain ranking is not an objective measure of pain. It's a proxy for an objective measure of pain. The closest we can get with existing tools. Just like the ATAR is not the student, and not necessarily representative of the students' abilities.

My Physiotherapist, Mark Scholes, told me in 2020 that he suspected that people's pain scales actually changed during the pandemic – that the stress, anxiety, and isolation of covid were causing people's perception of pain to change. Mark uses an evidence based approach to physiotherapy, measuring the force that can be applied by different muscle groups, and the angles people can achieve with various joints without causing pain. If Mark's suspicion is true (and I suspect it is) then a pain score of 4 pre-pandemic might have become a pain score of, say, 6 mid-pandemic, without any actual change in the underlying problem. Just to complicate matters, some folks' pain perception might have moved the other way (the average change is meaningless in the specific case), some might not have changed at all, and some might have intensified more than others.

We're back to the beguiling comfort, certainty, and illusion of objectivity provided by numbers – Luke Stark's "Charisma of numbers." We always risk forgetting that the measurement is not actually the thing we are measuring. Technology also brings with it the risk of tech solutionism – the idea that we can use technology to solve human problems. Sometimes we can, but often we miss the mark entirely. Cory Doctorow uses the example of contact tracing[100]. When covid19 hit, governments around the world rushed to put out contact tracing apps – a classic example of tech solutionism.

The trouble is that contact tracing is an exercise in trust and communication. It involves in-depth discussions with infected people about where they have been and who they have been in contact with – particularly difficult information to extract from people who may have violated lockdown rules and are concerned about getting into trouble for it.

Apps can't do that. The so-called contact tracing apps that were rushed into service actually measured whether two phones were within bluetooth range for a certain period of time. Not the same thing at all as whether two people were in a position to exchange viral particles. Even if you assume the apps worked as intended (many did not), it's entirely possible to be within bluetooth range and not be in a position to infect/ be infected (for example, being in adjoining rooms with no shared airflow, or being in neighbouring cars in slow moving traffic or while waiting at a drive through covid testing clinic). It is also possible to be in a position to infect/be infected without your phones being in range, if, for example, your phone was outside in the car, in a bag in another room, or on your desk while you amble down to the tea room.

We keep trying to reduce situations to quantitative, measurable data. Quantitative data can tell you what has happened, but it can't usually tell you why, or whether it was good or bad. Unfortunately, it does have a seductive certainty to it. You can use multiple decimal places to make data look incredibly precise, even where those decimal places are completely meaningless.

For example, the Bureau of Meteorology might tell me with great confidence that the temperature in my suburb is currently 23.1 degrees celsius, but that doesn't actually tell me the temperature in my garden, as temperature can vary widely within surprisingly small areas, depending on things like the amount of concrete, number and type of trees, and the current inland reach of the sea breeze. Another example is polling data that has an error margin of +/-3% but is reported to two decimal places.

Another risk of assigning a number to something is that the number can spread, even in the absence of evidence, and become an entrenched proxy for actual understanding. As we saw in Chapter 3 with the widely accepted 5 micron particle size that was assumed to be the only way particles could hang in the air for the length of time necessary for aerosol spread of a virus[101]. The 5 micron fallacy was so entrenched that studies showing that aerosols could be much larger could not get published. They simply weren't considered plausible. Much like Barry Marshall and Robin Warren's findings on helicobacter pylori causing stomach ulcers.

A similarly persistent number without foundation is the estimate that the internet generates 2.5 quintillion bytes of data every day. This appears to be a classic case of what Randall Monroe in his xkcd comic calls "citogenesis" – where everyone cites the number because everyone else cites the number, and the actual origins become obscured. The earliest mention I could find was an IBM blog post last updated in 2014[102], (though IBM tweeted the number in 2013, in a bid to sell more data storage) which gives absolutely no reference or method for producing the number. It appears to be a guess. Yet if you google "how much data is produced every day" this number appears all over the place.

Hard numbers offer a wholly deceptive, unwarranted confidence. They can persist and spread like a nasty infection. We need to turn to qualitative data to redefine what's meaningful.

The Unexpected World

In Chapter 2 we explored the way the world would look if our decisions were evidence and data driven rather than ideological. But not all of the examples in Chapter 2 were ideologically driven. Sometimes our decisions are based on the idea that we understand the world, when we actually don't, and that can derive from what really amounts to faulty pattern matching. In other words, we see something happening that we think we have seen before, so we react to it the same way we did last time.

For example, at the Supercomputing conference in 2016 I heard a keynote from IBM's Katharine Frase, who described a time when she shadowed a cardiologist on his daily rounds. She noticed that they had seen five patients on this day that were very, very similar. She notes: *"When the sixth patient walked in, what I observed was that the cardiologist replayed the script from the first five. Now, he may have been quite correct {but} to my layman's eyes, there were some significantly different things in that sixth chart to the first five."*

Frase admits that she does not know the correct diagnosis, but the issue of falsely identifying the next case as similar to the last one is a known problem in many areas of human endeavour. It's very easy to fall into a pattern and miss significant differences.

What if the way we teach actually encourages this kind of faulty pattern matching? If we are taught that we can predict the outcomes of every experiment, and produce answers that are obviously one hundred percent correct, then we are not teaching students to challenge their thinking and be sceptical of their results.

Real problems don't have guaranteed solutions. It's not always possible to tell if your solution is good, let alone if it's the best choice. Real problems require evaluation and experimentation. They require adaptability and creativity. We know that the real world does not behave in a reliable and predictable way, and few solutions work in every possible situation, or for every possible person. What if we could teach our kids that no solution is perfect, no scenario is predictable, and nothing is quite the way we expect it to be?

The Unexpected Classroom

Most of us grew up in an education system where teachers were the font of all wisdom. Keepers of perfect knowledge. I was lucky, in a sense, to be a teacher of Computer Science in a world of rapidly evolving technology. I had to come to grips very early on with the idea that there was no way I could know everything, and that my students would almost certainly know more than me in many different areas. It relieved me of the burden of expecting to be omniscient, and it meant I rapidly got used to saying "I don't know, let's find out!" I also learned to enjoy learning from my students, rather than seeing it as a sign of my own failure. Changes to our understanding of other subjects, say classroom maths, for example, don't happen at nearly the same rate as changes to the field of technology!

One way to teach technology and constrain the classroom to only the things the teacher knows, which I have seen some teachers embrace with worrying enthusiasm, is to have every student do exactly the same thing. Writing the same program in precisely the same way, building the same device, working in lock step. Though it's comforting, especially for teachers without strong technical skills, because it makes it much easier to diagnose errors and help students finish the work, this technique is doomed to failure from several angles.

Firstly, those students whose skills are already far in advance of this project are likely to become disengaged at best, sullen and disruptive at worst. (Remember Austin, my cautionary tale from the introduction? A classic example of a student who was not sufficiently challenged.)

Secondly, students whose skills are not up to this kind of task are also likely to become disengaged and/or sullen and disruptive, because not only is this a task they can't yet do, they will now feel bad about themselves for not being able to do what everyone else is doing. Pitching work too high or too low often produces similar outcomes in the students affected.

Thirdly, because there is zero room for creativity, requiring every student to do exactly the same thing in lockstep actively punishes students who think differently, who solve problems with creativity and innovation. Precisely the kind of skills we say we want our students to develop.

The good news is that technology also provides us with an excellent alternative to working in lockstep, allowing us to teach the skills and knowledge students will need to analyse and communicate data, while also building their critical thinking skills and emphasising creativity.

We'll go into detail with specific examples of these kinds of projects in Chapters 6 and 7, but for now let's consider what uncertainty looks like in the classroom. Once the teacher has let go of knowing the answer, class time becomes an exploration of ideas, and a chance to test out theories and develop a deeper understanding. Many excellent teachers already teach this way, of course, but the curriculum, for the most part, does not support it, because of the sheer amount of content that must be covered, especially in senior years.

Exploring ideas and having challenging discussions in the classroom is one thing, but how do we create projects that build in uncertainty? How do we support students when they get stuck on projects that have no clear solutions? How do we assess students' work when we can't label it right or wrong?

One of the highlights of working with real world datasets in my classroom was the discovery that they lend themselves beautifully to differentiation – the ability to create projects with different levels of challenge, for students at different levels of ability. The challenge, of course, with real world datasets is that you have no idea what the students will find. This is both an advantage and a disadvantage.

An advantage is that students can ask their own questions of the dataset, meaning that every student is doing a different project and exploring something they find uniquely interesting. The disadvantage is that the teacher cannot possibly do every project in advance, and understand every challenge the students will encounter. There is no list of right answers that they can check student work against.

This requires an entirely different approach to marking, where the process the students follow is the focus. As well as requiring the student to produce an answer to their question, you must also require the student to evaluate their own answer, and validate it in different ways. This means that they are able to tell you *why* they believe their answer is valid, as well as any reasons they might have to believe that it is invalid, which is a crucially important step we often neglect in the real world! This is, of course, an upside as well as a downside. If students routinely consider the possibility that their answer might not be correct, they are already streets ahead of many professional data scientists.

Setting up a project like this using an existing dataset is not as difficult as it sounds, as long as you understand the dataset well. This is one of the reasons I founded the Australian Data Science Education Institute – to find and make sense of datasets, to make them accessible for classroom use. Finding a dataset that hasn't been fully analysed is trivial – I could hit half a dozen with a bread roll from where I am sitting. But making sense of it is often challenging. Figuring out what each field means, how it was measured, and how it relates to the real world can be a lot of work.

The first step in exploring the dataset as a class is to make sure that everyone understands it. Which means you must start with the context. Why was the data collected? How was it collected? What are the limitations of the collection process? What (or who) is missing from the data? If it's sensor data, what are the strengths and limitations of the sensors used? Are they inaccurate at low concentrations, for example, or confused by unexpected inputs? I have an air quality meter which goes berserk every time I use hand sanitizer, warning me of dangerous levels of formaldehyde. My hand sanitiser does not contain formaldehyde, or anything that can form formaldehyde, but something in it confuses the sensor. If we were collecting that data, we'd have to understand that high formaldehyde readings could be erroneous.

As you work through the context of the data, you can also work through each field, exploring what it means, and how it relates to the real world context.

Then you need to consider what questions the dataset can answer. This is surprisingly challenging for many students. When we first worked with voting data from an election, the first question the students said it could answer was "which is the best party?" which led to a long and fascinating discussion about quantitative versus qualitative questions, how we define "best", and whether my definition of "best" is the same as yours. You can then move into a brainstorming session about different types of questions, considering which ones the dataset can answer and which it can't, and include a focus on how accurate or reliable the answers could be, given the limitations of the data.

For a different example, consider whale observation data, where the data collected is how many whales observers have seen breach the surface. The data that the scientists want is how many whales are actually passing through that stretch of water. What they *have* is how many they saw breach the surface, or actually, how many breaches they saw (they won't always know if the same whale breaches twice). So any calculations done on that data are, of necessity, estimates. We have a tendency to forget the uncertainty when we report results, reporting them as hard facts rather than as estimates based on uncertain data, so it's crucial to train students from the start to report the problems with their data, as well as with their calculations.

Once you've explored the kinds of questions that the dataset can answer, you can set students free to find their own question, and begin to explore the dataset to find an answer. Some students will need to have a question more or less handed to them on a plate, others will roam excitedly through the dataset, choosing and discarding questions moment by moment, before settling on a wildly challenging question that provides them with the motivation to strive for the moon.

At this point teachers can identify students working on problems that require similar technical skills, and spend some time working through those skills with the group, or with the whole class, in and around the students' individual work. Those who are already far ahead of those skills can work on more complex questions, while the others learn the basics.

All of this makes for a somewhat chaotic classroom, and while it might sound like a lot of work, it can actually be a very responsive way to teach that does not require detailed lesson plans for each session – indeed, you can't create such plans, because you won't necessarily know what help the students need in advance. Instead you have a series of micro-lessons in your toolkit that you can use for teaching particular skills, always using the current dataset as your example.

For the students who need their hands held, success might look like successfully solving a very small problem that they didn't think they could do. Simply teaching these students that technology is not terrifying, and definitely not beyond them, counts as a win.

For the students aiming for the moon, some will inevitably miss, and the marking of the project must be set up to reward the techniques used and the lessons learned, rather than whether the student hits their target or not. These types of students often struggle with not achieving what they identify as perfection, so this will be an incredibly important, and deeply challenging lesson for them.

Above all, teachers have to be prepared to have students doing things the teacher doesn't understand, and learning skills that the teacher has never learned. I have a PhD in Computer Science, and in every class I taught there were always students doing things that were outside my skillset. The trick here is to learn how to help the student find supportive resources, and to ask the student to explain their work – often this leads to the student finding their own errors as they explain their process to you.

Assessing the Unexpected

The simplest way to assess a student's knowledge is a multiple choice exam, because you can simply overlay a template and mark by rote, or, even better, you can feed it to a computer system to mark for you. No complexity. No ambiguity (at least in the marking). Very little effort. Real projects with no solutions must be at the other end of the scale for marking effort.

One of the reasons we use exams and toy assignments is that they're rather like the way we teach science experiments. They have fixed inputs, a known process, and fixed outputs, and it's very easy to tell whether they have worked or not. But if our definition of "worked" doesn't effectively include those qualities that we are looking for – that ethical, rational, critically thoughtful, creative problem solving – then maybe the fact that it's very easy to tell that it has worked is actually a trap from which we urgently need to extricate ourselves.

Assessing projects that don't have right answers is an entirely different problem. These projects require you to assess the process, to expect students to be able to explain and validate their answers, and to build in a whole range of skills that are often seen as peripheral to the set of facts and skills demanded by the curriculum, but without which the curriculum is quite meaningless. Skills like communication, critical thinking, and problem solving.

Rubrics then focus on questions like: Can the student explain the problem? Have they understood and explained the real world context? How thoroughly have they tested and critiqued their own solution? Have they clearly communicated the constraints and limitations they faced, with datasets or other aspects of the problem? Have they clearly explained the constraints and limitations of their solution? Have they understood and communicated the ethical implications of their work?

These rubrics are placing the emphasis on the skills we say we want students to leave school with, but that we rarely assess for. They are necessarily more subjective than we think a multiple choice test is, but as we saw in earlier chapters, multiple choice questions are biased in their own way, selecting for students who think in particular ways, and who interpret questions in the intended way. The maths example of a person who has $300 in their wallet and deposits it in their bank account, where the question is "is this transaction positive or negative?" rewards students who can guess what the question wants them to say – and I don't see that skill on the lists of graduate attributes bandied about by universities.

It is obviously difficult to assess this kind of problem solving & creativity in an exam, but let's think, for a moment, about the issues with exams. If exams are closed book, then they are, at worst, an exercise in recall – effectively fact regurgitation. If they include questions requiring creativity and problem solving, then they require those in the utterly artificial setting of no time to think, no access to information or resources, and no possibility of collaboration. Indeed, collaboration in an exam is considered cheating, whereas in the real world, effective collaboration allows us to achieve more than we ever could individually.

Exams are so popular because they are seen as an objective form of assessment that makes cheating difficult, if not impossible. They can even be blind marked so that teachers don't know whose work they are assessing. What could be better? Except… exams prioritise rote learning above all else. The only thing we can be confident of when someone gets high marks in exams is that they excel at getting high marks in exams. This would be wonderful if the real world only required us to get high marks in exams in order to solve problems.

One of the unexpected upsides of the pandemic is that exams had to move online, at least for a time, meaning they were open book by default. This seems to be a step in the right direction, except that, of course, it means someone else can write your answers for you, because there's no way to constrain communication. If, on the other hand, students are doing real tasks in real time as projects (as opposed to fake tasks under artificial time constraints), teachers can regularly check in and make sure the student understands what they are doing.

Getting comfortable with uncertainty is not a panacea for all of our educational woes, but it is an excellent start in allowing students the room to explore different solutions, and in training them to critically evaluate their own work. After all, if your answer is confirmed by the textbook solution, there's not much room for further evaluation. If we can use real problems without right answers as the basis of at least part of our education, then we will surely prepare our students for the real world.

So what, exactly, do these projects look like? How do we build them? Chapter 6 takes you through my journey – from teaching tech with toys, to using real projects. We'll look at the impacts on students' attitude and motivation, as well as some of the challenges I encountered along the way.

6

Projects with Impact

In the Beginning, there were toys

When I first started secondary teaching, at a new Science School, I was teaching two subjects: A year 10 Introduction to Computer Science subject called Creative Studies that had been created by a Computer Science academic, and a year 11 Computer Science subject that I had the freedom to create the way we wanted it with my co-teacher, Victor Rajewski.

Creative Studies in its first year, before I started, included Robotics, Simulation, "Natural Computing", and Python programming. The students used a system called StarLogo, a block based programming language similar to Scratch, which is what's known as an agent-based simulation system. This means that the code students wrote was to control "agents", or "sprites" on the screen, that could then interact with each other, and with their environment.

MIT's StarLogo TNG programming software

The simulation project in that first year, before I started teaching at the school, was bushfires. The students could take it as far as they wanted, creating houses, trees, different terrain, fire fighters, home owners, etc. So it was a real system they were simulating, but the scenarios were fake, because StarLogo was not nearly complex enough to simulate any real scenario.

There were several problems with that course. Of the 8 classes taking the course, only one was taught by a teacher with actual programming experience, which made it exceptionally difficult for the students to get help when they encountered problems with their code. This is a distressingly common situation in schools, perhaps because anyone with programming skills can earn a lot more in almost any other industry, than in teaching!

There were also a lot of things that StarLogo simply could not do. Any student attempting to program anything complex would run up against the limitations of the system pretty quickly, such as limits on the number of agents that could exist at any given time, complexity of the terrain, etc.

There was also a fairly severe lack of documentation and sample code available online – often an issue with programming languages designed for teaching, simply because they lack the large cohort of programmers sharing tips and tricks online that languages used in industry tend to collect. What there was online, however, was a lot of sample bushfire code. Many students dealt with the frustration of trying to do things the system couldn't do by simply downloading the sample bushfire code and submitting it as their own. Given the lack of support available, and the frustration of using that particular system, this isn't terribly surprising, and it wasn't detected at the time.

The Robotics section of the course used Lego Mindstorms NXT 2 robots. While Lego is famous for its easy to use, straightforward building instructions, its Mindstorms system is anything but simple. Students had to program the robots to follow lines, push each other out of circles, or find a particular coloured section on a map. The maps were huge and cost a fortune to print. The robots themselves were expensive. And the students hated the course.

There were endless problems with loose wires, flat batteries, and broken sensors to overcome, not to mention the difficulties of learning to program in Mindstorms itself. Some students, who often had Mindstorms robots at home, blitzed these challenges and were bored by the lack of creative choice. Others were frustrated and dispirited by the projects, unable to see the point of wrestling seemingly insurmountable difficulties to achieve something with zero impact. Once again there was a desperate lack of teachers who could program to help the students. Many students learned that programming was both something they hated and could not do.

The Natural Programming section of the course used slime mold, with the aim of showing how slime mold will find the shortest path to food through a maze. Once again, the lack of teachers with experience – in this case experience with slime mold – meant that many of the experiments failed, but there were no teachers qualified to explore why the experiments had failed, so it was another frustrating experience.

The final section of the course was Python programming, where students wrote programs to play 21, and tic tac toe. Now they were using a real programming language (the third they had been expected to learn that year – StarLogo, Mindstorms, and now Python), but again they lacked teacher support, with only one teacher who knew Python available to 220 students.

Creative Studies was a classic example of the 'toy' approach to teaching Computer Science. Even though Simulation and Robotics can be serious, practical topics, in Creative Studies they were taught with toys – StarLogo and Lego Robots. And even though Python is not a toy, the Python programming section focused on games – 21 and tic tac toe. There was nothing in this subject to convince these science-minded, intelligent students that programming had anything to offer them.

Creative Studies was hands down the most hated subject in the school. The students hated it. They could not see any connection between what they were learning (or unable to learn) in that course, and anything they would want to do in the future. And this was a science school where most of the students wanted to go into a science career, where they would almost certainly need programming. The teachers also hated it, because they felt out of their depth with assignments they could not do themselves, and because they were constantly battling the students to try to persuade them to do work.

The Real Thing

The year after Creative Studies made its unpopular debut, Victor Rajewski and I started teaching our year 11 Computer Science subject. We made it an overview of Computer Science, including Artificial Intelligence, Ethics, Algorithms and Data Structures, Computational Science, and Usability. For the first half of the year we discussed a lot of complex ideas, such as the nature of intelligence, and built foundational programming skills in Python.

In the second half of the year, while still covering more content in class, the students worked on a Computational Science assignment where we worked with scientists in order to meet their computational needs. A lot of scientists have little or no background in programming, which significantly limits their research. If you can't use programming as a tool to manage your data, you have to do things like finding duplicates, picking out values that match a particular criteria, and analysing and visualising your data, by hand. It can take a lot longer, and in some cases be a lot less accurate. Because spreadsheet packages such as Excel have a hard limit on the number of rows of data they will load (Excel's current limit is around one million), some datasets simply can't be processed at all without programming.

Lack of data and computational skills can lead to other issues for scientists too, such as the recent problem in the UK where covid case numbers went missing because they hit the column limit in Excel (just over 16 thousand columns) and the data was silently truncated or thrown away[103].

In that first year, we worked with a Marine Biologist, Dr Kathryn Wheatley, and a Molecular Biologist, Dr Michael Gantier. Kathryn had a bunch of tracking data about seals, and Michael was doing cancer research and looking for relationships between different types of microRNA. Kathryn was hoping to find patterns in the seal data that would make searching through the reams of data easier by identifying sections in the accelerometer data that indicated the seals had caught a fish, for example, and making it possible for her to zero in on that section of the data. Michael needed to compare different RNA samples with results from a range of databases – a task that took weeks to do manually.

Both projects were complex, and involved working with real data. I made sure that the researchers knew up front that, since this was a student project, there was no guarantee that there would be usable results, and the students knew that their marks for the assignment did not hinge on producing something that actually made a difference to the researcher's work.

Instead, their marks were based on documenting the problem, producing a functional, usable, and clearly documented piece of software, and identifying the limitations and errors in their own work. This last step was crucial, and became a cornerstone of my approach to projects in the future. I have worked with so many senior, skilled people who struggle with the idea that their own work is not perfect. This renders people unable to improve their work, and it also makes it exceptionally difficult to collaborate. To find the best solutions, we have to be able to improve on the ones that we have. That's impossible to do without recognising that what we have is not perfect. The capacity to identify flaws in your own work is crucial to developing effective solutions to real problems.

The seal project turned out to be much, much harder than I had anticipated. Part of a learning experience that continued for several years, it helped me zero in on what kinds of projects worked best, but at the expense of the students, who weren't able to achieve high levels of success and actually produce work that supported the researcher. Interestingly, though, the students still seemed to count it as a success. They did well in terms of marks for the assignment, but most of all, they learnt a lot about working with real data, about programming, about working in groups, and, of course, about seals. It seemed that doing something real, even if it was very difficult, and in some cases out of their reach, was far more engaging than playing with toys.

I'll talk more later in this chapter about what I learned about the kinds of projects that are most successful. For now, I want to talk about the cancer research with Michael Gantier. One group who tackled this project consisted of two students, Chris and Matt. Both students had significant programming skills coming into the subject. In addition to his strong programming skills, Matt had studied Biology and understood what micro-RNA was, and some of what Michael was trying to do. Chris had no Biology experience but very strong programming skills.

Michael Gantier, meanwhile, had enough Computer Science background to know what he needed, but not enough programming expertise (or time) to write the code himself. He was spending weeks on tasks that he knew could be automated, so when he saw my call for projects, which I had sent out through various university academics I knew, he jumped at the chance.

For several months, Chris and Matt worked closely with Michael to produce a piece of software that allowed him to search various genetic databases and find matches in a matter of hours, rather than days. Michael was thrilled with the software, and went on to use it in his research. Chris and Matt could honestly say they had done cancer research in year 11, and knew that they had produced something of value.

As a typical software client, Michael then started to fizz with enthusiasm over what else could be done, and this is where something happened that really emphasised the value of doing real projects: Chris and Matt continued working on the software over the summer break – long after any possibility of receiving course credit had passed. They continued working on it during their year 12 year as well, optimising the software so that it completed searches in minutes rather than hours, and adding in extra features that Michael wanted.

I asked Chris in 2020 – 9 years after he studied my year 11 subject – what he remembered from that project, and this is what he said. *"Although I started your unit already with interest and experience in the technical side of programming, the class was my first encounter with the realities of real-world data. In particular, real research data presents challenges in its un-*

certainty, opacity and magnitude that cannot be appreciated without hands-on experience. This early exposure equipped me with an understanding of what it means to pull meaning out of data, as well as a readiness for when I would perform analysis on my own results later in my research career." Chris Whittle.

You might say that this project was a fluke. That these were two high performing, extremely motivated students who got excited about a project. And that would certainly be a fair description of Matt and Chris, both of whom are currently working on PhDs. But it kept happening. From molecular biology to neuroscience. From Physics to Conservation Ecology. Every year at least a few students would get so excited by their projects that they'd keep working on them the next year.

Remember my cautionary tale from the introduction, Austin? He was wildly disengaged with the year 10 "toy" approach to Computer Science. We couldn't get him to do any work. I was not looking forward to working with him in the year 11 subject. Austin, though, turned into one of the best students I ever had. He partnered with another student, Michael, for the computational science assignment, and together they produced some of the most extraordinary work I have ever seen.

The project that year sprang from a talk I went to at the Nepean Conservation Society. The speaker that day was Judith Muir, owner of Polperro Dolphin Swims, a dolphin touring company in Port Phillip Bay that is dedicated to the protection of the local dolphin population, and the overall health of the marine environment. Judith talked about the data they have collected about the dolphins over the years, and how that data gets sent to the government regulators, where, as far as she can tell, it sits on a shelf somewhere. Of course, the combination of data and dolphins was irresistible to me, so I waylaid Judith after her talk, and we quickly became collaborators and co-conspirators.

Judith gave me access to the data, which was handwritten notes from each tour over a period of about 8 years. I found volunteers to transcribe it, with all of the usual challenges of handwriting, format, and interpretation that handwritten data brings with it. I had students produce heat maps of where the dolphins were typically found, among many other different ways of wrangling the data, but Austin and Michael took a different approach. They decided this hand written data thing was a nonsense they could do without, so they decided to create an app that the Polperro crew could use to record the data, automating the entry of common phrases, and including gps coordinates rather than the location estimates that they had been using, which included "East of Point King" and "Just off Whitecliffs".

They produced a problem specification document that was one of the most professional I've seen, in schools or in industry. The problem specification is the part most programming nerds loathe, because they see it as tedious paperwork, and it was rarely a highlight of these assignments. As well as the app they produced a server side program that would run on the Polperro's office computer, to produce the forms required by the government regulators from the information collected by the app. The whole project had a focus on reliability and usability that is rarely seen in student projects (though it was a focus of the marking scheme). Austin & Michael continued working on the project over the summer and into the next year. Remember, Austin didn't do a stitch of work with the toys, but, together with Michael, he produced an outstanding project as soon as I gave him something real to do.

That project was amazing, not merely because I got a few extra trips on Polperro as a result (the Polperro crew are the most passionate, knowledgeable, and skilful team, and a trip on Polperro is an extraordinary delight), but also because it confirmed for me the motivational power of doing something meaningful.

In the year before I left teaching, 2017, one of the projects was another genetics based data project with Dr Sonika Tyagi, from Monash. This time there was a machine learning aspect to the work. There were three projects that year, and the scientists involved came and pitched the projects to the students, who then chose the one they found most interesting, or most challenging. I was rather startled to be approached by a year 11 student not studying Computer Science by the name of Viktor, who pleaded with me to be allowed to work on the genetics project. He laid out his credentials in Biology and Programming in great detail, assured me he was excellent at working in groups, and promised me that he would be able to make seriously useful contributions to the project. How could I say no? That project produced such strong results that the students involved presented their work at the Lorne Genome conference at the start of 2018.

When students consistently kept working on the projects after the assignment was over, when students who weren't even doing the subject asked to do the assignment... that told me this was something special.

It really confirmed to me that data projects needed to spread. That, while I was having a significant impact on my own students, every Australian student needed access to projects this engaging, this motivating, and this real.

Getting Real

With projects this successful and motivating in year 11, I was increasingly convinced we were on the wrong track with our year 10 subject. From 2010, the year before I started there, until 2016, we continued with variations on the same theme. We used different programming environments (but always block based), and tweaked the projects, but the problems were always the same, and the feedback from the students was terrible. The Head of Faculty who was in charge of the subject was convinced that the subject worked, though, and was reluctant to change it.

One of the things that startled me about working in a school was that there was no culture of evaluating courses or teaching. It was common to write a new course, run it, and never ask the students for feedback on the course, nor systematically track their results. It was one of the things that frustrated me when I was trying to show that the course we were teaching was not working, because it was very difficult to get permission to run feedback surveys.

Eventually the faculty head did run a feedback survey, but it was not anonymous. Of the 200 students taking the subject that year, we had 20 respondents, all aware their identities were known, and all saying anodyne, safe things about the subject. The questions on that survey were leading, too. It is well understood in research that asking students to agree with the statement "The subject was fun and engaging" will get a very different response to "How did you feel about the subject". How you ask the questions on your survey makes a significant difference to your results.

Because we were now in a pitched battle about how the subject should look the next year, I eventually managed to persuade the school to run some focus groups about the subject, led by a teacher from a different faculty. Now the students were able to give feedback to a teacher who was not connected to the subject, and wow! Were they honest!

The feedback was direct and to the point. The fundamental message, which underpins all of the issues with the toy approach to teaching technology, was that the students could not see the point of what they were supposed to be learning. And kids can be quite pragmatic – when they can't see any point to what they are doing, they don't do it.

We had highly capable students failing the subject entirely because they refused to do the work (the official response to this was sometimes to force the student to repeat the subject – with predictable results). We had other students cheating, copying each other's work, copying work off the internet, or simply paying someone else to do the work. And we had many, many students who tried, because they wanted the marks, but wound up demoralised and miserable, convinced that they hated computing and that computing hated them.

All we were succeeding in doing was proving to the students that computing was a nightmare, and in most cases something they were not equipped to understand. The exact opposite of what we needed to do.

After that crisis point, there was still a lot of back and forth about how the subject should look, but I eventually got permission to trial a small project. I had one student who was absolutely enthralled with politics, and the strange way preferences flow in the Senate for the Australian Federal election, so since we had recently had an election, I trawled the Australian Electoral Commission (AEC) website to look for data we could use.

I hit the jackpot, in the form of a comma separated values (csv) file that contained every single Victorian vote for the Senate in the Federal Election. It was a treasure trove, but like all real datasets, it was quite messy. For a student project, though, that was perfect.

Each row of the file looked something like this:

Aston, Bayswater, 1,1,1, 1,,2,,,,,,,,,,3,,,4,,,,,,,,,,,,5,,,,,,,,,6,,,,,,,,,,,,,,,
,,,

The first field was the name of the electorate, the second was the polling place, the next three numbers were some kind of internal data we didn't bother with, and the final string of numbers and commas was an encoding of the actual ballot paper[104], where each comma denoted a new square.

The ballot paper looked like this:

Senate Ballot Paper Excerpt for Victoria from the 2016
Australian Federal Election

Except that it went on for 40 columns, with a total of 151 squares on the full paper. The complete ballot paper looked like this:

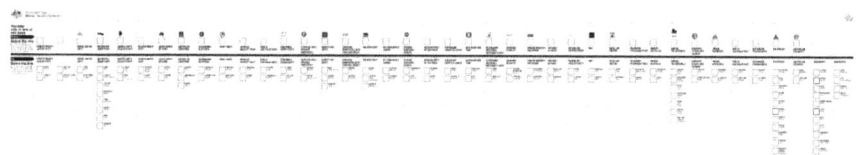

Full Ballot Paper, not shown actual size!

The first step in dealing with this dataset was to figure out how the one dimensional string of numbers and commas representing a vote mapped on to the two dimensional ballot paper. This was further complicated by the fact that the columns on the ballot paper could contain anything from 2 to 12 entries. I called the AEC, who were unable to find me anyone who could explain the mapping. Presumably they *have* such people, but the people on the switchboard either could not or would not find them for me. So I spent some time writing Python programs to figure out which people got the most votes, and finally deduced the mapping using the reported results and the results of my Python scripts.

To understand the true magic of this project, we need to know a little about Senate voting[105] in Australian Federal Elections. In the 2016 election the rules had changed. People could put at least 6 numbers above the line, starting from 1, voting for whole parties, or they could put a minimum of 12 numbers below the line, again starting from 1. Each number is only allowed to be used once – so you can only vote 1 for a single party, OR a single person. You can't vote above and below the line, you must choose one or the other. If that's as clear as mud, these are the AEC's own words:

On the white Senate ballot paper, you need to either:

- number at least six boxes above the line for the parties or groups of your choice, or
- number at least 12 boxes below the line for individual candidates of your choice.

In dealing with the ballot paper, the students naturally assumed that when they processed the string representing each vote (over 3 million of them), they would only find each number once, that people would either have voted above the line or below the line but not both, and generally that voters would have followed the rules. Of course, people are not very good at following the rules, plus the rules had just changed, and the file was a faithful representation of exactly what was written on every ballot paper, meaning that some lines contained the number 1 several times, some voted above and below the line at the same time, and some just did wild things that made no sense according to the rules. It was the perfect introduction to Data Science as an authentic study – no real dataset is entirely predictable and perfect. You have to assume that your dataset will contain the unexpected.

We spent some time as a class discussing what types of questions the data could answer, and then the students got to devise their own questions to explore. The questions included

- "How many people voted against the rules and hence did not have their vote counted?"
- "Which party's voters were more likely to follow their party's "How to vote" card?"
- How did the voters in *my* electorate vote?
- How did the voters at *my* polling booth vote, and was that different to my electorate as a whole?
- If people voted 1 for Party A, who did they vote 2 for?
- Where did the people who voted for Party B come from (eg city, suburban, rural)
- What percentage of votes did women candidates receive, compared with the percentage of candidates who were women?

I taught the students just enough Python to extract the answers to their questions, and then they created engaging visualisations to answer their questions. These were not simple graphs, though the scale needed to be accurate, but creative images that were more engaging and compelling than a graph could be.

Once again, a student took the project and ran with it, extending it into a preference flow prediction tool that he was still working on at university some years later. That was exciting, but what really impressed me was the turnaround in feedback for the subject overall. By this time I was running anonymous feedback surveys at the end of every unit, and now the vast majority of students were saying that they could see the point of the work, that they were using the skills learned in that unit elsewhere, and that the subject was absolutely relevant to them and their futures.

Sadly I was not able to bring work samples or survey results with me when I left the school, but the impact of that shift was so dramatic that it has stayed with me, a constant motivation for my work at the Australian Data Science Education Institute.

The next year we used data collected by local volunteers about microbats in Melbourne with one group, and climate change data with another, and both projects again saw high levels of engagement and motivation from the students.

With the microbats we did a small sample project initially where all of the students analysed the data for an answer to the same question – Which batbox, or nesting box, orientation did the bats prefer in the different seasons? This was an incredibly simple set of data with a very easy question to answer, and I expected the visualisations to be 48 variations on the same theme – most likely something involving a compass with different sized bats on it. I was astounded when the assignments were submitted. The range of creativity and innovation that the students had used in representing this simple dataset was phenomenal. That was a pivotal moment to me – the data did not need to be complex or challenging in order to engage the students. It just had to be real and meaningful.

Though the batbox orientation data was simple – just a count of how many bats had been found in each box in each season, and the orientation of the boxes – the full dataset for the microbats was quite messy, as the volunteers had put multiple sets of information in single sheets, changed the labels they used for some values, and occasionally changed the way they recorded their data. Again, a messy and authentic dataset was just what the students needed to learn real Data Science skills, as opposed to how to answer textbook questions on a perfect dataset, and how to pass exams.

The Climate Change dataset we worked with was even simpler. It was a set of temperature predictions based on a range of different climate models, predicting average global temperatures from 2015 to 2090. All the students had to do was extract the data from their chosen model from the file, graph it, and create a final visualization to make the data compelling.

Some students did far more than that, seeking out historical temperature data from other sources so that they could create a visualization showing the changes over a much greater timescale – a much more compelling image. Some worked with predicted sea level rises, creating visualisations that showed playgrounds underwater, or a child's bed afloat, or the barrels of salt that would represent increasing salinity of warmer oceans. Some just did the basics of the assignment, but then went on to use the skills they had learned in their Science projects and elsewhere. They marvelled repeatedly at how useful and meaningful these skills were.

As always, some of the students needed a lot of help to find their question and decide on a compelling visualisation. The subject had a lot of problems, too. We did not have anywhere near enough skilled teachers to teach every class, and one of our skilled teachers became ill, leaving us even more short staffed. The students were not supported enough when writing their code and analysing their data. Yet, despite that, the final feedback survey showed that the vast majority of students had a sense of achievement, and felt that the subject was deeply relevant to them and their futures.

It's important to note that what we were teaching was not what a professional Data Scientist would recognise as Data Science. It was really basic data literacy, very simple data analysis, and visualisation as communication, more than as a complex technical skill. But these are the essential foundations of the discipline, which are often glossed over in the rush to technical skills. If a data scientist does not understand the context, the limitations, and the flaws of the data they are working with, and does not know how to communicate their results effectively to others, then all the complex technical skills in the world will not make them effective at their job.

If all the students took from the subject was the ability to critically analyze graphs, to look for the limitations and flaws of a dataset, and to challenge and critically evaluate their own work, they'd still have been far better equipped to understand the Covid19 pandemic, as well as to challenge the ideological rhetoric that too often takes the place of evidence based policy making in both business and governments.

I've described some of the projects I taught, but how would you go about creating your own project along these lines? In my work with ADSEI, I have devised templates to make the creation of these projects easier, and I'm building a curated set of datasets to take the work out of finding & making sense of real world datasets. In Chapter 7 we will look at the templates, and some more examples of projects that you can do across all year levels, right across the curriculum.

7

Templates for Data Science Projects

Local Projects

People often assume that Data Science in Schools has to be secondary school only, because how could primary kids do Data Science? The truth is that Data Literacy and Analysis skills can be built into the curriculum from as young as 5 years old. And it's really important that kids learn Data and Tech skills early, because by the time they get to secondary school we've already lost a lot of them. They believe that these skills are too hard, not relevant to them, or just not interesting. We need to show them early on that Data Science is a useful tool that they are more than capable of mastering.

So how can primary kids do Data Science? Like any other Data Science project, it's crucial to put it in context, so the kids can see the point. These local projects can be used in Secondary Schools as well, where the kids find a problem in their school, or in their local community.

The outline of these projects is simple:

1. Find a problem the kids care about
2. Measure the problem
3. Analyse the measurements
4. Communicate your results
5. Propose a solution
6. Implement your solution
7. Measure again to see how well it worked

STEP ONE: FIND A PROBLEM THE KIDS CARE ABOUT

It might be litter in the playground, traffic at pickup time (or, to put it in a way kids will really relate to – how long they have to wait to be picked up, or how far they have to walk to the car!), or access to play equipment. It might be shelter for homeless people, access to fresh food, sharing home grown fruit, vegetables, and herbs, or raising money for a charity that resonates with them. You can also do awareness raising projects, where the kids come up with ways to educate the school community about a global issue, as well as ways to measure effective awareness both before and after the project.

Sometimes people fret that not all kids will be enthusiastic about the same problems, and this is absolutely true. This is, of course, also true about any given assignment. The interesting thing that I have found is that it doesn't really matter what the problem is, as long as it's real. I can't guarantee you that you will engage every kid with every project, of course. Some kids are hard to reach at times, and sometimes that has more to do with what's going on in the rest of their lives than what's going on in any particular classroom. But I *can* assure you that overall engagement is higher as soon as the project is real.

I must add a little caution here about Design Thinking projects that say kids will find solutions to problems, but never actually implement them. These projects might look fancy and engaging – Let's plan a trip to Mars! Let's cure world hunger or solve famines in Africa! They are chock full of wonderful media moments. But they are dispiriting for kids for several reasons.

We promise them they get to develop solutions, but they never get to actually implement the solutions and see if they work. This might explain a lot about politics these days, as it reinforces the idea that solutions are perfect and complete in themselves, rather than what they actually need to be in the real world: iterative, constantly monitored and evaluated, and always able to be improved.

It's also quite dispiriting for kids to get all engaged with an issue and not be able to take any steps towards fixing it. Kids have big hearts, and while they obviously need to learn that they can't fix everything (something I still struggle with some days!), they can also be empowered to create positive change right here in the present.

STEP TWO: MEASURE THE PROBLEM

For this step to work, it is essential for students to understand the problem, and to know what it is they are trying to change. This means they need to come up with a metric that is somehow measurable – whether by direct observation, online surveys, or monitoring behaviour. For example: Count and identify the litter, time how long people have to wait to be picked up, measure how far people have to walk to the car, or count the number of people who get to use the monkey bars every lunchtime for a week.

One project I co-designed with staff at Gillen Primary School in 2020 looks at the school community's knowledge of local indigenous languages.[106] The metric in this project is how many words students know, on average.

Students survey each class and total up the results, calculating the average number of words known per student. Then they design and implement a school wide awareness raising campaign – it might be a word of the day for the whole school, or regular events, competitions... the key thing is that the students design it themselves.

After running their campaign for some pre-determined amount of time, the students run the survey again, and figure out how well their idea worked. Which classes improved the most? Which did not improve? Did any go backwards? They could then improve on those ideas, or come up with totally new ones, and run the whole project again.

STEP THREE: ANALYSE THE MEASUREMENTS

Analysing the measurements can involve anything from very simple maths to complex statistics. For younger kids, that might simply mean sorting the rubbish into categories (eg chip packets, icy pole wrappers from the canteen, and sandwich bags or cling wrap from home), or organising the drop off or play equipment measurements by year level or by day. For older kids you might enter it into a spreadsheet and use a formula to calculate some averages over the week or by area or year level.

In this step it's also crucially important to explicitly explore the ways in which the measurements were flawed. For example, was the litter collected on a windy day, or during a week when one year level was away on camp? Was it after an event? What other factors might influence the amount of rubbish in the school grounds?

For survey data, it's always important to consider whether everyone who filled in the survey did it accurately, and come up with reasons why they might not. For example, on surveys of eating habits, people typically report that they eat more fruit and vegetables and less junk food than they actually do.

Pawsey Supercomputing Centre and ADSEI have a sleep survey online[107] where people sometimes report that they do not have a mobile phone AND that they sleep with their mobile phone next to the bed, and in an earlier version of the survey that had a write-in box for the country you slept in last night, one person reported that they had spent the preceding night on Mars. Sometimes people simply hit the wrong buttons or misunderstand the questions, sometimes people are actively messing with your data.

If you collect data about traffic by counting the number of cars, there are any number of ways the data could be flawed: It could be an atypical day, such as a rostered day off at a nearby building site, or a university holiday, or a student free day at a nearby school. I once saw our local roads authority measure the traffic on a major arterial road near a university during university holidays when the traffic was much lighter than usual. They were surprised when the measures they implemented wreaked havoc during normal traffic conditions.

The people counting the cars might have over or under counted (it's surprisingly difficult to count passing cars accurately, dealing with lapses of attention, remembering how many of the white cars you just saw you actually wrote down, among many other issues). The weather might also be a factor (less people walk or catch public transport if it's raining).

Regardless of how you collect your data, there will be flaws in it. It's very important that students learn this from the start, so that when they meet datasets or reported results as adults, they will know to ask questions about the origins of the data, and to recognise that no dataset is perfect. To actively incorporate a discussion of the flaws into every data project is an incredibly powerful way to build data literacy and rational scepticism.

Once the data is analysed, students can check their analyses with each other to verify their results, and they can explore other ways of analysing the data. Sometimes we see the same data interpreted to mean wildly different things, so it's important to explore ways each students' analysis might be wrong, or misleading, or incomplete. It's also important that the students recognise that there is no single way to analyse any given dataset, and no single right answer to their problem.

STEP FOUR: COMMUNICATE YOUR RESULTS

It's no use finding things out if you can't communicate them. This is where you graph or visualise your results. The littlies can "graph" the results by stacking up blocks to represent the different categories. Green blocks for chip packets, blue ones for icy pole wrappers, etc. This is a great, tangible, exercise in data representation. Older kids can draw graphs or do them in a spreadsheet like Excel or Google Sheets. It helps to get them to draw pictures and labels on their graphs to make them more interesting and compelling.

There are important questions the students need to explore at this stage, too. What are they trying to communicate? What message do they want their audience to take away from the visualisation? Have they conveyed the values in a way that is misleading in any way? (For example, by not having elements of the graph sized in proportion to their value, or by not including zero on the scale.) Is their visualisation interesting and compelling? Is it clearly labelled, and are the labels accurate?

For example, if you were graphing the total litter picked up in the schoolyard, kids might want to label it "total litter dropped". But litter picked up is only a proxy for litter dropped. Some of the litter might have blown away, or not been found because it was under some play equipment. Some of the litter might have been picked up by others.

Or, when graphing the results of different approaches to the litter problem, you might be tempted to title the graph "The best way to solve our litter problem." In actual fact, what you have is a few approaches, and which one led to the least litter being found in the grounds. There might be better ways, or there might be reasons why less litter was found that aren't related to the approaches tried. This is a great time to teach the kids that any data representation that claims to answer a qualitative question is suspect.

STEP FIVE: PROPOSE A SOLUTION

Now the students devise possible solutions to their problem. For litter they might come up with nude food day campaigns, or a change to the way food is available in the canteen – such as using larger chip packets and handing out small paper bags with chips in them, instead of lots of small plastic packets.

For traffic it might be that pickup times can be staggered by year levels, or older kids might be encouraged to walk further and be picked up a block or two away. The students might propose walking school buses (groups walking together along a common route), or run a walk-or-ride-to-school campaign.

This is a great time to compare solutions and discuss the pros and cons of each idea, emphasising that there is no such thing as a perfect solution. You can get kids to brainstorm what might happen as a result of each idea, and propose different possible outcomes, so that they understand that they can't be sure what will happen without actually trying their solution *and measuring the results.*

Using the information they have, students pick the solution they like best, and plan how long it needs to run in order to show measurable results. Sometimes it won't be obvious how long it will take to show results, in which case it might be necessary to plan for several different possibilities.

STEP 6: IMPLEMENT YOUR SOLUTION

This can be a whole school initiative, and involves a lot of communication, using the graphs from Step Four to tell the community what's happening and why. This step takes time, depending on the project, so you might set a program in motion and let it run for a few weeks to be sure it has time to have an impact.

It's important, for this stage, to have the class consider what other factors might influence the program. Are some classes away on camp or on excursion? Are there changes to local traffic flow for other reasons? Has the weather been unusual? If, for example, you are trying to change electricity use, has it been unusually hot or cold, or unusually cloudy so more lights have been on? Or has there been more computer use for some reason, compared with the time when you took your first measurements?

Sometimes even professional Data Scientists forget that they are not doing their experiments under controlled conditions, and that any changes they detect might have reasons other than the ones they are looking for. It is crucial that we teach students from the start to look for reasons why their projects and experiments might not be working as expected, and why their results might not mean what they expect or want them to mean.

STEP 7: MEASURE AGAIN TO SEE HOW WELL IT WORKED

This is my favourite step, often sadly missing from political initiatives. Once you've tried to fix something, you need to measure it again to see if you actually made any difference.

You can even repeat the process with several different solutions to compare which ones work better.

There are several key questions to ask at this stage:

- What's different between this set of measurements and the first set?
- What conditions have changed, apart from the things we tried to do?
- What other reasons might there be for the differences we see in the measurements?

- How could we change the way we measure to make it more accurate or more meaningful?
- If we're not sure that the difference we see in the measurements is because of what we did, how could we test it?

It's very important to challenge the students' certainty at this point, and be sure that they have identified all of the reasons why they can't be 100% confident of their results. This is a lesson we need to emphasise as part of every project they do, to ensure that they are rationally sceptical, not just of other people's data and results, but of their own.

So at this point the students have identified a problem, measured it, tried to solve it, and measured how well their solution works. They've also critically evaluated their measurements and their solutions. That's more rigorous than most government programs, and, indeed, most educational reforms.

I love this template because it is the essence of STEM – It's a science experiment, devised by the kids, with rigorous measurement and evaluation. Maths and Technology are used in handling the data, and you can use Engineering to design your solution, or to measure the problem. For example, you can use sensors if you're looking at environmental conditions like heat, noise, or water.

You can scale the technology use up or down depending on available resources and where your students are up to. There are no robots with parts to fail. And the best part is that the motivation is built in. The kids are learning that STEM and Data Science are tools you can use to solve real problems in your community. They're not just a bit of "fun" that isn't all that much fun, and is not relevant to their futures.

There is a sample Local Project at the end of the book.

Global Projects

In contrast to the local projects, global projects use existing datasets to explore global issues. The challenge here is to figure out what questions the datasets can (and can't) answer, while maintaining the critical focus that examines the way the data was collected, stored, and analysed in order to identify possible shortcomings.

This type of project can be done with everything from beginner spreadsheet skills to advanced programming, which makes them wonderful opportunities to create projects that extend every student in your class *just the right amount*. I've never met a class of kids yet that does not have students with an extraordinary range in the incoming level of technical skills.

What they *never* have, though, is an understanding of how to do Data Science. Even at the Science school I taught at, where most of the students were high achieving and highly motivated, they had never been taught how to critically analyse a graph. They did not know how to ask simple questions to explore the validity of a graph – questions like:

- Why doesn't the Y axis start at 0?
- What was the sample size of the data?
- How was the data collected?
- What population was the sample drawn from?
- Who/what was missing from the data collection?
- Who/what is misrepresented by the data?
- What were the limitations of the sensors that recorded this data?
- What other factors might have interfered with the collection of this data?
- Is this analysis/interpretation of the data valid?
- What other interpretations could this data support?

So, we start with a dataset. It might be about climate change, air pollution, deforestation, income inequality, or the average happiness of different countries and the factors that contribute to that. It might be about politics, diversity in the tech industry, people's understanding of science, homelessness, housing affordability, or taxation. It might be about something more light hearted like pet ownership, or something more intense like the health gap experienced by many indigenous populations. It might be about microbats, rainfall, turtles, or dolphins. It might be about careers, or sport, or public toilets.

There doesn't seem to be any limit to the interesting datasets that are out there just begging to be analysed.

STEP 1: MAKE A COPY

Step 1 is always to make a copy of your dataset before you start playing with it. In these marvellous days of automagical version histories, where software allows you to see previous versions of all of your documents, you might not need it, but it's a good habit to get into – to make sure you're not altering your original dataset.

A friend of mine who is a statistician tells the story of researchers doing a food study consisting of a survey with around 120 questions. They were asking things like "How many times a week do you eat broccoli?" and, since some people didn't answer all of the questions, they assumed that no answer was the same as 0. Unfortunately there are many reasons for leaving a question blank, ranging from accidentally skipping it, to feeling embarrassed to say you eat chocolate at every meal, so my statistician friend pointed out that making all of those blanks 0 was not a valid way to handle the data. "Let's just leave them blank," she said. Unfortunately the researchers had already turned all of the blanks to zeroes, and had no way to revert to the original data. They no longer knew which values were genuinely zero, and which were blanks that had been converted to zero. The study was largely invalidated.

STEP 2: UNDERSTAND THE DATASET

This is a step that differentiates the way we do Data Science from the approach to data in the maths curriculum. Maths teaches us to graph column A against column B using, say, a scatter plot. Data Science the way ADSEI teaches it requires students to understand the dataset before they can figure out which columns to graph, and how to graph them.

So here we ask basic questions like:

- How many columns are there?
- How many rows are there?
- What does each row represent?
- What does each of the columns mean?
- How was the data collected?
- What is the range of each column? (what are the highest and lowest values)
- What are the units for the values in each column?

Despite the textbook approach to datasets that simply requires column X to be graphed against column Y, independent of their meaning and context, it is essential to teach students that no dataset can be analysed without first understanding what it represents.

If we understand the context, we are better able to determine what values make sense to graph against each other, and what questions the data can answer – and therefore how to analyse the data. If we teach Data Science as a matter of blindly applying formulaic techniques to every dataset, then we encourage a kind of mindless reliance on numbers, with no way of questioning their validity or meaning.

STEP 3: LIST THE PROBLEMS WITH THE DATASET

No dataset is perfect. Sometimes it's proxy data for what you actually want to know (for example, counting the number of cars in a carpark when you actually want to know how many people came in those cars – given how many cars carry only the driver these days it's probably a reasonable approximation, but it's not the same). Or it's data that excludes some portion of the population (for example, a count of the number of people coming out of the station using the stairs, which must miss anyone using a wheelchair and coming out using the lift), there are always flaws.

One dataset I worked with consisted of concentrations of different pollutants in the air, and sometimes the values went negative. It's not possible to have less than none of something in the air, so negative numbers don't make a lot of sense. It turns out that the particular sensors that dataset used got kind of wonky at low concentrations, which meant not only that the negative values should have been zero, or very close to it, but also that the very low values were not particularly accurate.

A great way to start this step is to have a class discussion brainstorming possible flaws. Make a list of questions that need to be answered – for example, details of the sensors used, or of the data collection method – and then students can spend some time trying to find answers to those questions. It won't always be possible, but that just means there are more possibilities for flaws in the dataset that should be noted.

One of the reasons for making the flaws explicit – and including them in any final report – whether they are speculative or known, is that it also makes doubts about the final analysis explicit. You will rarely see this kind of approach taken with any reported results, but it is an important part of combatting the "charisma of numbers".

STEP 4: EXPLORING THE DATASET

Exploring the dataset is a great chance to build simple spreadsheet skills. It's amazing what you can learn from a dataset simply by sorting, filtering, calculating sums and averages, and graphing different fields against each other. This is also a great time to emphasise the difference between *Median* and *Mean*, which is something I never properly understood myself until I started to teach with real datasets.

The *mean* is the average of a set of numbers, so you add up the numbers, divide them by how many of them there are, and that's your mean. The mean can be shifted quite a lot by outliers in your data – so if you have a whole lot of data between, say, 50 and 100, a single value in the thousands can shift your mean quite a lot. The *median*, by contrast, is the middle value in a dataset. Much less susceptible to outliers. This is particularly relevant for datasets like income, or property prices, where single values can be dramatically different to all of the others.

For example, compare these two datasets representing annual wages in thousands (so 80 represents $80,000 per year):

										Mean	Median
80	80.5	70	62.5	97.2	55	68	74	59		60.8	70.7	69
80	80.5	70	62.5	97.2	55	68	74	59		1000	164.6272

You can see that a single outlier in the second dataset (an annual salary of $1,000,000, not an unusual salary for the CEO of a large company, but out of reach of most of us) has more than doubled the mean, whereas the median has only shifted by 3. If you're using mean to talk about salaries, then the salaries of people like Bill Gates and Jeff Bezos make the "average" salary look pretty good by skewing the mean high. It's like measuring the weight of your fruit & vegetables while leaning on the scale. Terribly misleading. Whereas if you use median, you get a much better sense of where the middle of the data really lies.

This is also a useful opportunity to explore the weaknesses of the values known as Summary Statistics. It's very tempting to characterise a dataset using the mean, correlation, and standard deviation. Anscombe's quartet is a classic example of four datasets with nearly identical summary statistics, but wildly different distributions[108]. It's a classic cautionary tale: Don't assume that the data is the same, simply because the summary statistics are the same.

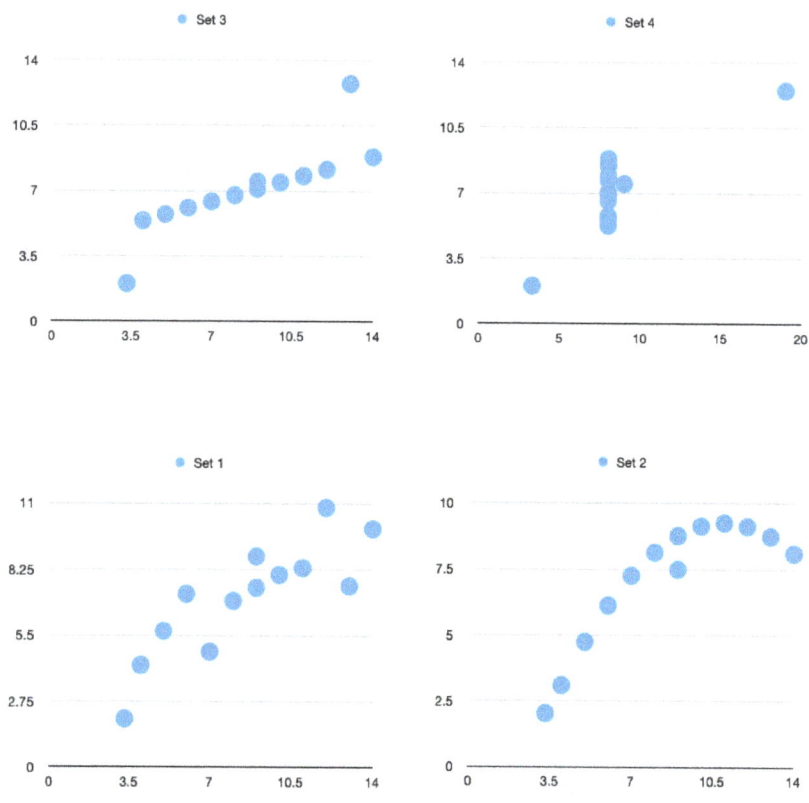

Anscombe's Quartet has almost identical summary stats but wildly different distributions

Now is also the time to emphasise the need for rational scepticism and double checking of results. Some time ago I found a really interesting dataset, showing small scale solar installations by postcode over time, around Australia[109]. I decided to write a lesson plan around it, and the first thing I did was sort the data to get a list of the top 20 postcodes for solar installation around Australia. I was startled to find that, for the year I was looking at, 18 of the top 20 postcodes were in Western Australia and Queensland. I did not expect those two states to have the best approach to renewable energy, so I started writing up the results in great excitement.

To provide extension activities for those students who can code, I decided to write a quick program in Python to figure out the average by state (in Australia it's easy to determine the state from the postcode, as it is coded by the first digit. So 3102 is in Victoria, 2102 is in NSW, etc). When I looked at the results, I found that, actually, WA and Qld had the lowest solar installation average. I checked and rechecked the results, and on closer inspection it turned out that, while a few postcodes in each state had high numbers of solar installations due to large new housing estates (which have to have solar to meet the Federal Government's Sustainable Energy requirements), for the most part solar installations in those states were quite low.

At first I was embarrassed – I had made a rookie mistake, gone against my own teaching, and failed to double check and verify my initial findings. How embarrassing! But then I went and looked at the media headlines surrounding that dataset, and do you know what they all said, without exception? *"WA and Qld the best for Solar!"* Seems like rookie mistakes are more common than critical thinking and careful analysis.

Graphing is an important way of understanding a dataset. Human beings are much better at detecting patterns in graphs than in lists of numbers. But it's very important to distinguish between graphing for understanding and graphing for communication. At this point you'll graph lots of things just to see if any trends or patterns jump out at you. When you graph later for communication, choice of graph, labelling, scales, and a whole lot of other considerations will be key. At this point, we're really just playing to see what we can find.

Graphs can show you trends and patterns, as well as the key points where things change, but they can also make outliers very obvious. It is sometimes tempting to simply delete outliers. We often speak of outliers as being so rare as to be irrelevant, but depending on the nature of the outlier, and on the population size and selection criteria, among other things, outliers can be very important to your data.

For example, some years ago I did a sleep study that involved recording when I went to bed every night, and when I woke up. The recording site used 24 hour time, which is not the way I am used to thinking about time, so the first few nights I accidentally said I went to sleep at 10am. You can't necessarily simply throw values like that away, though, because what if someone is a shift worker, or stayed up all night at a party?

Similarly, in a medical study, if you throw away the one outlier who reacted badly to a drug in a population of 100 test subjects, when the drug is subsequently taken by hundreds of thousands of patients, 1 in 100 can quickly become thousands of patients adversely affected. Sometimes the outlier is actually the value you are looking for.

STEP 5: FIND A QUESTION THE DATASET CAN ANSWER

Before you analyse your dataset, you need to figure out what you're looking for. What questions can this dataset answer? When I first used voting data with my year 10s, they all said that the data could tell us which was the best party. This was an understandable answer, but it quickly leads to problems. What do you mean by "best"? Who gets to define it? Data can't, in general, answer qualitative questions like which is the best or worst, why did people vote that way, or which is the nicest.

What data *can* tell us is which party got the most votes. It would be nice if that correlated with which party was the best, but sadly even if we can agree on a definition of best, there are many more factors influencing the way we vote.

So what we can get from the data is answers to quantitative questions. Questions like "which party got the most votes, which one sold the most, which value is highest, which building is tallest" etc.

Sometimes we frame quantitative data as a qualitative question, like "when is air pollution the worst?" but those questions actually boil down to quantitative values like "when are the particulates in the air highest?"

You can have some wonderful classroom conversations about the difference between quantitative and qualitative questions, and what the data can actually tell you, versus what you'd like it to be able to tell you.

This doesn't mean that qualitative data is useless. Far from it. Qualitative data can tell you the why, the how, and even what we can do about it. It's just a lot more work to analyse qualitative data, and you rarely get a nice clear number out the end. I would love to build Data Science projects where students collect qualitative data and learn to analyse it, but that's probably a whole new book. One giant leap at a time!

STEP 6: FIND THE ANSWER TO YOUR QUESTION

It's very important to recognise that the answer to your question might be no answer at all.

One of the most common Data Science traps is to show correlation and use it to mean causation. For a closer look at this trap, do a search for Spurious Correlations. My favourite is the Spurious Correlations website, which graphs things like divorce rates against margarine consumption[110].

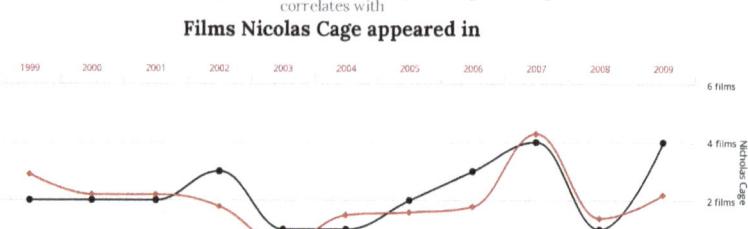

Spurious Correlations Chart from Tyler Vigen: Swimming pool drownings vs films with Nicholas Cage

If students are asking whether one thing causes another, there will very likely not be enough information to answer that definitively. The best you can do is to show that two things are correlated. But just because thing A is correlated with thing B, it does not mean that thing A causes thing B. We have a tendency, as human beings, to interpret correlation as causation. This is because, in many situations, that assumption works, and may, in fact, save our lives. If Fred ate berries and then died, it doesn't cost us much to avoid eating those berries, but it might well prevent us from suffering the same fate.

In more modern examples, you could use the fact that death rates from the flu went down in Australia during the 2020 covid19 pandemic to say that covid19 prevents the flu. What actually happened, of course, is that the lockdowns, social distancing, masks, and hand washing designed to prevent the transmission of covid19 also prevented the transmission of the flu.

To actually prove causation requires careful and methodical experimental design which is not always possible or even ethical. (For example, you can't selectively infect people with things to see what happens!)

This means that you need to be very careful of what you say you have found in your data. For example, if you are considering the WAMSI whale data, you can draw conclusions about how the numbers of whales sighted off the coast has changed, but not directly about how the number of whales have changed (it's that proxy data issue again)[111].

The more we teach our kids to report on what they have actually found, rather than what they think it means, the better chance we have of a future where people are careful and realistic with their results and assumptions.

Many interesting questions can be answered with very simple forms of data analysis, easily conducted using basic spreadsheet functions. If the only technical skills students acquire are how to calculate sums, averages, medians, and how to filter out values that match a certain criteria, as well as how to create valid and meaningful graphs, we would still be producing students with much higher technical skills than most of us have as adults.

For teachers and students with coding skills, more sophisticated questions can be answered more easily. It is possible, for example, to answer the question "for those who voted 1 for the LNP in the senate, who did they vote 2 for?" in a spreadsheet, but not easy (and most spreadsheet packages won't even open a 3 million line file). In code it is much easier to extract that kind of information. This is a great way to motivate students to learn how to code, and languages such as Python and R make this kind of data analysis very accessible, in quite small programs.

STEP 7: CHECK YOUR ANSWER

We often skip this step in assignment work, not to mention in the Data Science industry, but it is crucial to teach students to be sceptical of their answers. Find different ways to calculate their result and see if they work. Get multiple students to calculate answers to the question. It's a start to give them all the same process and see if they get the same results, but much better to have them independently come up with the calculations they will need in order to answer the question, just in case the process itself is wrong.

It's useful at this step to have students ask themselves what other interpretations could apply to their results. We have a terrible tendency to assume that our first interpretation is both correct AND the only possible interpretation, so building this practice in to every assignment is a great way to make sure that we teach open minded scepticism.

You can get students to check over each other's work and ask each other critical questions, to build the habit of challenging results rather than accepting them as valid from the moment they appear on our screens. As they become more familiar with challenging the work of others, you can build the habit of them challenging their own results, as well as building their ability to take constructive criticism and use it to improve their work.

No answer should ever be submitted without a full accounting of the work done to verify the validity of the result, *and* a discussion of all the ways the result might be invalid or misleading.

STEP 8: COMMUNICATE YOUR ANSWER

The most amazing result in the world might as well be no result if you can't communicate it effectively. Communicating results involving data nearly always involves some form of graph. Unfortunately the closest most maths curricula get to teaching data communication and visualization is to teach students how to draw particular graph types for any given dataset, completely divorced from why you might want to teach that graph type. Most students know that linear data goes in a line graph and discrete data in a bar chart, but that's where it ends.

This was actually one of the things that spurred me to fight to be allowed to change our year 10 course to a Data Science course. All of the students did an Extended Experimental Investigation project over quite a large chunk of year 10, and almost without exception, the graphs they used to display their results were wildly inappropriate, and often completely invalid.

The best graph type to use is not only a function of the type of data (though that's important), it also depends on what your question is, and what aspect of the data you most want to highlight. For example, in recent years it has become fashionable to dismiss pie charts as a bad way to represent data, and often this is true. But there are times when it is very effective to represent your data that way. If, for example, you wanted to show the percentage of the Australian population that are fully vaccinated as at May 29th 2021, a pie chart is a very compelling representation of how poorly vaccinated we are as a nation.[112]

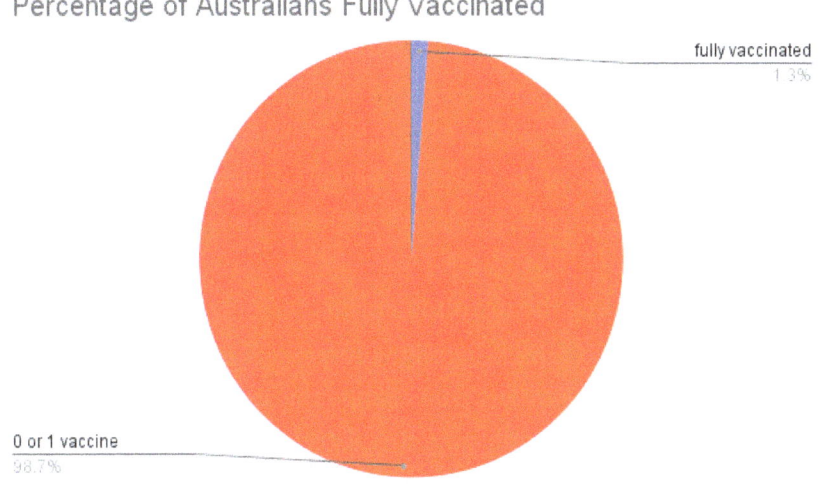

Percentage of Australians fully vaccinated against Covid19 as at May 29 2021

It's a poor representation if you want to compare two or more similar values, or to read off the size of a particular segment, but to get a sense of how far we have to go, it's quite effective. Of course, this chart is misleading because it compares fully vaccinated with everyone else. Many Australians have now had one of the two injections needed for vaccination, so we should really include that. Unfortunately that information is quite hard to find, as the government is being less than transparent with its data.

Questions you can ask when trying to choose a graph include:

What do I want people to understand from this graph?

As with any form of communication, this is a key question. What am I trying to say? How easy will it be for people to understand the key points from this graph?

What do I want people to be able to read off this graph?

If they need to be able to read particular values off the graph, would grid lines help? Is each data point labeled with its value? Is it obvious which line, bar, or point is which?

How can I make the graph maximally readable?

Sometimes this means reducing the amount of data you are displaying on one graph, by graphing a subset of the data, or aggregating the data so that one point on the graph represents multiple data points. Both of these techniques can also be used to mislead, though, so you must also ask:

How can I ensure the graph is valid?

Does my choice of graph highlight some data in a misleading way? Would including more or less data change the apparent meaning of the graph? Is the trend line indicating a real trend? Who is your audience and will they understand this type of graph?

What is the right scale to use for this graph?

The scale is a place where many graphs mislead, either deliberately or accidentally. For example, using a log scale is a scientifically valid way to represent datasets with a very large range of values, as you can get more data onto a smaller graph. But it gives a very different impression of the rate of change, unless you are very used to reading log graphs.

Consider these two graphs of covid cases detected in the world, over time. They both show exactly the same data, and they are both scientifically valid representations of that data. The first graph uses a linear scale, where the vertical axis increases by a constant amount, so each tick on the axis is an increase of 50 million. The second graph uses a log scale, where every tick on the vertical axis is 10 times larger than the tick before. So it goes from 10 to 100 to 1000, etc.

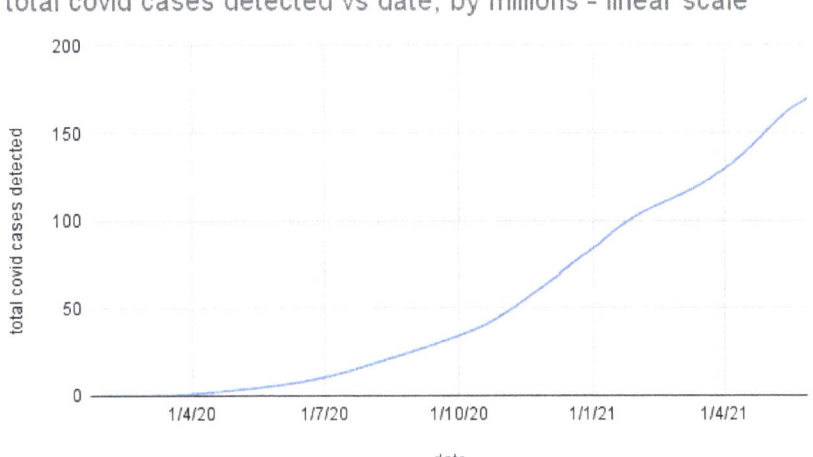

Linear graph of worldwide detected covid cases over time

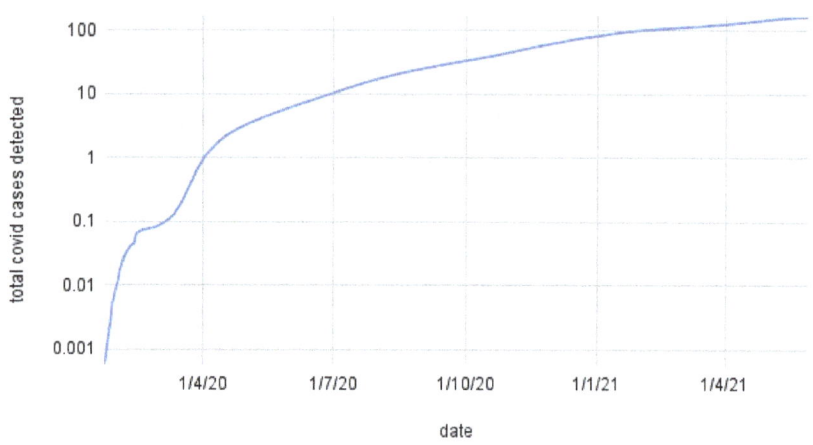

total covid cases detected vs date, by millions - log scale

Log graph of worldwide detected covid cases over time

The rate of change on the linear graph shows that cases are still growing. The log graph shows exactly the same thing, except that people who are not familiar with log graphs might think that the cases are levelling off, when they are not.

It's clearly not enough to be able to say that a graph is valid. As with any form of communication, you must ask: will my audience be able to make sense of this? Will it mislead or confuse them in some way? How can I avoid that?

MEASURING IMPACT

As with the local projects, it's possible to measure impact for the global projects by using the data to raise awareness of global issues. You can do pre and post surveys to see how well the awareness raising worked, and you can even run competitions for the most effective ways to communicate the data.

There is a sample Global Project at the end of the book.

Where to from here?

This is by no means a comprehensive description of all of the possible data projects you can create, or every last detail of how to create them, but it gives an overview of what the project looks like, and what kind of skills students will learn as they tackle authentic projects using real data. The messiness and complexity of real datasets leads to a richness of learning that you just can't find using a faked dataset.

I once tried to build a fake set of weather observations simply to include sample graphs in a weather forecasting project I wrote for Primary School kids, and it turns out that faking data in a realistic way is much, much harder than it looks. That's a good thing, because it makes fake data easier to spot. Any data that is perfectly regular or has a perfect curve or nice neat summary statistics is instantly suspect. And it's a great thing, because it means that students won't learn their Data Science skills in a nice neat sandbox where the answers are known and everything works on the first try. They'll learn *real* Data Science skills, including how to handle complex, messy, confusing data. More than that, they'll learn that Data Science skills are tools they can use to create positive change in the real world.

So where to from here? How do we move on from our existing education system, with its focus on exams and "facts", and upgrade to a system that prioritizes critical thinking, innovation, and ethics? Chapter 8 lays out the next steps to take us to an education system that prepares us for the future.

8

What Now?

When the new Australian Digital Technologies curriculum was created in 2015, each year's work built sensibly on the year before. Unfortunately the decision was taken to bring the entire curriculum into the classroom all at once, meaning teachers were trying to implement a year 8 curriculum for which the students had not done the preparatory work from years 1-7. But that was only part of the problem.

By far the bigger problem was that teachers were now being expected to teach a curriculum that they themselves had never been taught, either at school or at University. Many secondary Digital Technologies teachers are not qualified in the discipline, and Primary School teachers are still not being trained to teach Digital Technologies, even though it is a mandatory part of the curriculum.

This leads to a somewhat circular problem. Teachers haven't learned this way, or been taught to teach this way, so how can we expect them to teach it? Clearly we need to change the way we teach today, *and* the curriculum, *and* teacher training. And do to all of that, we need to change what we expect and value in education.

Education has really failed to evolve. Sure, we've tweaked around the edges, but fundamentally my kids are still experiencing classrooms the way I did, the way my parents did, the way their parents did, and on down the ages. We don't use corporal punishment anymore, and kids are allowed to move around a little more, and make some more noise, but when you get right down to it, students are still sitting around waiting for teachers to tell them the facts they need to remember, and the exercises they need to do. And, above all, they're waiting for the tick from the red pen that says they got the latest exercise right.

The pinnacle of school success is still getting a hundred percent on the exam. It's coming out with a 99.95 ATAR, getting into Law or Medicine, being Dux of the school, or topping the class. It's clear that we have a system that perpetuates socioeconomic inequality *and* prioritises all the wrong things.

What if, instead, the pinnacle of school success was making life better for homeless people in your area? Massively reducing the amount of litter that gets washed into the storm water drains? Giving kids somewhere amazing to play? Reducing your school's energy use? Contributing to cancer research? Persuading your community to get vaccinated? Reducing the spread of a disease in a pandemic?

What if you got to learn by doing things that have the potential to make a difference in the world?

Sure, you say, sounds dandy, but is it actually realistic? Well, we know that what we're doing isn't working. I did a quick call out on social media to ask what the highlight of school was, and it was devastating how many responded with: *leaving*. A lot of people loved opportunities to sing in choirs and play in bands. A few people remembered their friends. You know what nobody said? *The exams. The assignments. All the facts we memorised. My exam results.*

What I'm proposing is not rocket science. Much of what I've described in this book has been known for decades as Problem or Inquiry Based Learning[113]. We know it works. We know it's engaging. I've just added Data Science and solving real world problems to a formula that we already know is effective. We know what to do. We know how to do it. We know why it will work.

So. Why don't we try new things? How do we shift the astonishingly conservative world of education onto a new track? This is not a technical problem, or even really an education problem. It's a political problem.

It's difficult to take a system that everyone has experienced for themselves and persuade the world that we need to change it. It's also difficult to persuade a political system that benefits from people being uncritical and not causing trouble to put funding and effort into changing that system into one that produces heretics. But a system that produces heretics is exactly what we need.

So how do we change it?

First of all, we need to give our best teachers autonomy. We need to remove the constraints of a content-stuffed curriculum that leaves no room for exploration, and we need to empower teachers to give their students the freedom to find problems that they care about.

When I was teaching I was forced to give my students a paper based exam, because "that's how we assess at year 11." We need to remove the requirement for particular types of assessment, and trust teachers to find ways of assessing their students' work that fits with the types of work they're doing, and that prioritises the things we want the students to learn.

So, if we're going to try new things, what's the one thing we absolutely need to do? We need to evaluate them.

We should try all kinds of new approaches to education, and we should systematically monitor and evaluate them. We should allow schools to try new things, to stretch their wings and fly. We should remove the pointless administrative burden from teachers and instead support them to find new ways to engage and inspire their students. But we should also carefully and rigorously measure the impact.

If we're going to measure the impact of new education systems, we need to be careful to consider the populations they're working with, and measure what business likes to call "value add". If we have a school that selects only the top students and then gets the top results, it might not actually be anything that school has done, rather that it is working with a group of very motivated kids.

Fortunately, it's not so hard to measure these things. We know how to take a scientific approach to educational evaluation. It's time consuming, and can be expensive, but how much more expensive than producing another generation of undereducated, disengaged students who come out of school believing they can't do STEM, can't do Maths, and aren't smart? Compare that with the possibility of nurturing a generation of students who rightly believe that they can (and should) change the world. Clearly a possibility worth investing in.

Like a Universal Basic Income, investing in education has been shown to produce society-wide benefits, as well as direct economic improvements. Studies vary wildly on how much the economy gains from investment in education, but the lowest estimate I could find suggested that every dollar invested in Early Childhood Education returns two dollars to the economy. That's the kind of return I will happily invest in!

What can you do?

If you're a teacher, you can use projects like the ones on the ADSEI website to teach the existing curriculum in a meaningful way. Our projects tick off curriculum goals for multiple subjects, so you can run with them as far as you like. And you can bring us in to help you develop community based projects that engage your whole school.

If you're a parent, you can share the ADSEI projects with your kids' teachers, and you can support the teachers who are already doing amazing work within the constraints of the current system. Above all, you can lobby education departments and governments for change.

If you're a policy maker, you can get support schools to do real projects and assess the things that matter, and you can support the development of curriculum that gives space for it to happen.

If you're someone who believes in a rational, evidence based future, you can share this book, have conversations about the parts you agree with AND, of course, the parts you don't. Give the book to everyone you can think of! And you can support ADSEI's work at https://www.givenow.com.au/adsei

My ultimate goal is to put ADSEI out of business, because teachers are teaching this way, teacher training programs are building it into their pedagogical practice, schools are producing kids who are changing the world, and governments are supporting and encouraging these outcomes. It might take a while, but we can do this, if you – all of you! – will support it.

As noted in Chapter 1, *"It turns out that diversity in the technology industry is only a small part of the reason why teaching all kids Data Science and STEM skills matters. The big part is that we need a technology & data literate population who are trained to think critically and creatively, and, in particular, trained to believe that they can solve problems. That's the world we need to build. And the foundation stone of world building has to be education."*

9

Sample Projects

Local Project

This is an example of a project pitched at Grade 3/4 students.

HOW MUCH ADVERTISING DO YOU SEE?

The advertising industry spends billions of dollars every year on trying to change our behaviour. Making us buy things, go places, eat things, watch things, use things, and so much more. How much advertising do you see in a day, a week, or a year? What kinds of things do you see advertised most often? If those ads work, what might the impact be on the things you do and buy?

YEAR LEVELS

Years 3 & 4

GOALS

In this unit the student will improve their data literacy and data analysis skills by :

Knowing:

- What data is and how it is collected
- That data can help tell a story about the world around us and help us solve problems

Doing:

- Collecting data
- Applying basic mathematical skills to make sense of data
- Sharing the data story with others in different ways.

Understanding:

- How advertising can influence people's feelings to buying things
- That data can be interpreted in different ways by different people for different reasons
- How communication is important when it comes to advertising

STEP 1

- Watch this Behind The News Episode on junk food advertising https://www.abc.net.au/btn/classroom/kids-ads/10538062
- And this one about sneaky ads: https://www.abc.net.au/btn/classroom/sneaky-ads/10540568
- In an online session, brainstorm to create a list of what kinds of advertising you see and where you see it. What makes something an ad? What are ads for? As the teacher, share your screen while you co-create the list with the class.
- Try to think of a whole lot of different types of advertising, not just ads on tv or youtube. For example, cars with company advertising on them, signs at the side of the road, t-shirts with logos on them, product placement in tv shows (for example where someone drinks Coke or Pepsi rather than just a cola, or where the apple symbol is visible when someone uses a laptop), posters in the supermarket, and others.

STEP 2
- Create a personal list of all of the types of advertising you can remember seeing.
- Estimate how often you see each type of ad – for example 5 times a day on youtube, once a day on tv, twice a day on billboards, once a week in the supermarket, etc.

STEP 3
- For a week, write down every time you see some advertising. Make a note of what is being advertised, and how. For example: food with pictures of tasty food, a holiday destination with pictures of people having fun, an antiseptic spray with warnings that you'll get sick without it, etc.
    - Try to categorise the type of advertising, for example:
        - Humour – getting your attention by making you laugh,
        - Being cool – making you want to buy something because people look cool using it, or because some popular celebrity likes it,
        - Fear – you need this product to protect you or your family from disease or danger
        - Health – you need this product to be healthy
        - What other techniques can you think of?
- Before you start this phase, plan how you will record and structure the information, given that you want to be able to pull all of the information together in the next phase of the project.
    - You will need to be able to count types of ad (billboard, online video, poster in shop, etc) and company or product advertised. For example, if you simply write a sentence every time – "saw a billboard advertising McDonalds with a pic-

ture of a burger and fries" – that will be difficult to count compared to using a table to collect the information.
- You also want to be able to make a quick note rather than writing a lot every time. Is there a way you can setup your notebook so that you can tick boxes wherever possible, rather than writing a lot?
- Your table might look something like this:

Date	Company	Video or Picture?	Location (eg youtube, tiktok, tv, billboard, magazine, etc)	Images (what kind of pictures are in the ad?)	Type of appeal (eg health, safety, hunger, belonging, status, etc)
5/5/20	McDonalds	Video	TV	Tasty Food, happy people	Hunger, belonging, happiness

STEP 4

- Work out how many types of advertising you noted down during the week that weren't on the personal list that you wrote before you started to count the ads..
- Put this information together as a class: how many types of ad did the class expect to see, and how many did they actually see? You could use SurveyMonkey or Google Forms to collect this data, and then share the results with the class. For kids without internet access, they could send the data by text or relay it over the phone to a friend who has internet.

- How difficult was it to put this information together?
    - Did you all use the same descriptions for the different types of ad?
    - Was it sometimes difficult to work out whether a type of ad you saw was the same as the type of ad your classmates saw? How did you solve this?
    - What might have happened if you couldn't easily discuss it, because the kids who collected that information were in another class or something?

    Step 5
- Make a bar graph of the number of types of ad the class expected to see, compared with the number of types of ad the class did see. How big does the difference look? Kids can do this online and share a document with you, or on paper and take a photo, or mail you the graph on paper.

STEP 6

You can break your data up in a lot of different ways. Make 3 different graphs using the same data by graphing things like

- Different categories of ad – eg food, entertainment, cars, clothing, etc
- Different companies advertised – eg McDonalds, Coke, Target, etc
- Different ad media – eg Billboards, Youtube, TV, etc
- Different techniques used in the ads – (see Step 3)
- See if you can think of other things to graph in this dataset.

STEP 7

Consider the different advertising techniques.

- As a class, discuss how they make you feel. Which ones do you think are more likely to persuade you to buy or use something?
- Which ones do you think are more likely to persuade your parents to buy or use something?
- Can you take the same techniques and use them to make your graphs from step 5 more interesting?
    - For example, using pictures to make them funny, or scary, or to make you think some things are cool?

STEP 8

Write a report for your family and friends, using the difference between how much advertising you thought you were seeing and how much advertising you actually recorded when you were looking at it, to warn them how much advertising really surrounds them.

Before you start writing, ask yourself what it is you really want people to know? Which of your graphs will you use in your report, to help get that message across? Do you need a different graph? Or do you want to make your graphs more interesting and compelling with pictures?

ADVANCED ACTIVITIES

Given the number of ads you saw over one week, how many ads would you see in a month, or a year? Remember that a month is usually not exactly 4 weeks. How could you account for this in your calculations?

Might there be times of the year when you see more ads than you saw in this week, or less? For example when you spend more time out and about, or in shopping centres, or online?

Is it possible that you didn't notice some of the ads you saw? Devise a new experiment that would make your results more accurate – ie make it more likely that you notice ALL of the ads that you see in a week.

Sample Global Project
SCIENCE REPORTING QUALITY

Science is often reported in the media. Clickbait headlines such as these sell more newspapers than sober reporting of the facts.

INSECTS 3 days ago
Giant spider hauls away man's pet goldfish for dinner

Media headline: Giant Spider Hauls Away Man's Pet Goldfish for Dinner

Near star could soon explode

SUPER NOVA Astronomers have become obsessed with a nearby star after it dimmed to its faintest level, potentially pointing to an imminent explosion.

Media headline: Near star could soon explode

Headlines shout about cures for diseases after preliminary results in mice show tiny successes. We are warned about the health risks of everything from food to mobile phones on the basis of studies that were of tiny sample sizes or had no statistically significant results. So how can we evaluate the quality of science journalism and separate fact from clickbait? This project will explore ways to identify high quality journalism. You'll study a range of different articles and videos and find ways of communicating their quality (or lack thereof) to your friends and family. Once you've done all of that, you'll evaluate publications, websites, or particular journalists for the overall quality of their science.

YEAR LEVELS

Years 9 and 10

SUCCESS CRITERIA

In this unit the student will improve their data literacy and data analysis skills by :

Knowing:

- What data is and how it is collected
- That data can help tell a story about the world around us and help us solve problems
  Doing:
- Learning to evaluate and identify the quality of science journalism
- Developing graphing and information visualisation skills
- Developing science enquiry, critical thinking, and communication skills
- Applying basic mathematical skills to make sense of data
- Learn the graphing skills that enable these comparisons
- Sharing the data story with others in different ways
  Understanding:

- How to check your data for possible errors
- How data and language can be bias and incorrect reporting or analysis can influence public opinion

RESOURCES

https://www.smc.org.au/for-media/tips-on-reporting-science
https://www.theguardian.com/science/blog/2012/mar/07/scientists-help-improve-science-journalism
Article used in supporting video:
https://www.sciencedaily.com/releases/2020/02/200227144259.htm

STEP 1

- As a class, brainstorm the characteristics of a high quality science experiment. Not class experiments where you know the expected outcome, but experiments designed to test a hypothesis. What does a high quality experiment look like? Does it aim to prove the hypothesis, or disprove it? Create a list of things to look for when evaluating a science experiment.

STEP 2

- Having identified the characteristics of a high quality science experiment, how would you write it up in order to highlight those characteristics? What would you look for in the write up to show that the experiment has been rigorous and unbiased?

STEP 3

Science journalism is one step removed from the write up of the experiment – it's a write up of a write up. What does this mean for quality journalism? What features of the scientific write up need to be reflected in a popular science article about the work?

Read this article, which reports on a study of the impact of caloric restriction on cellular ageing in rats: https://www.sciencedaily.com/releases/2020/02/200227144259.htm

Discuss the reporting of the article as a group, identifying any issues you see with the headline and summary, and then watch this video critiquing the article:

Think about the purpose of science journalism. It aims to help people to understand the research: what did they do, and why does it matter? Understand the implications of the research: what does it mean for us? What might it mean in the future? Understand the limitations of the research: what does it *not* tell us? (for example, work done in mice or other animals will not necessarily show the same results in humans – check out @justsaysinmice for examples).

- As a class, create a list of features to look for in a popular science article to show that it correctly reflects the work reported by the scientists, and the state of the science. This is your High Quality list.

STEP 4

As well as looking for positive features, sometimes negative features of a piece show up the quality of science reporting. What kind of features suggest that a particular piece of reporting is sensationalised, or otherwise poorly reported? Some things to consider are how does a particular way of reporting something change how we feel about it?

Read this article on relative versus absolute risk, paying close attention to the examples the author gives. How do the different ways of expressing risk make you feel?

https://www.healthnewsreview.org/toolkit/tips-for-understanding-studies/absolute-vs-relative-risk/

- Create a list of features that indicate low quality science journalism. Start with clickbait headlines that promise near-magical cures/solutions, and work from there. This will be your Low Quality list.

STEP 5

- Create a table, or other organised way to display your information and make a checklist using both High Quality and Low Quality to determine the quality of any particular article or video. As a class, work out where you will draw the line. Which, and how many, characteristics from the High Quality list must an article have to be considered High Quality? Which, and how many, characteristics from the Low Quality list must an article *not* have, to be considered Low Quality?

STEP 6

- In pairs, find 4 popular science pieces – articles or videos. Articles in newspapers or magazines, or on news websites are perfect, or youtube explainers about recent science results. For each piece record the writer/creator, the publisher (which magazine or newspaper, which youtube channel, etc), and the medium (video or text).
- When you've read/watched each piece, rate it for quality, just on your first impressions.
- Now go use your checklist/table to rate each piece according to your criteria.

STEP 7

- Compare your first impressions with the results when you used your criteria. How good a match were they?

- As a class, discuss your findings. Did your first impressions match your criteria? Were any of the pieces you evaluated wrongly classified by your criteria? For example, do you feel like high quality pieces, when evaluated using your criteria, were labelled low, or low quality pieces labelled high? What, if anything, went wrong?
- Refine your criteria, if necessary, to make them more effective.

STEP 8

Visualisation is a form of communication. To visualise data, you need to consider the fundamental principles of communication: What am I trying to say, and who am I trying to say it to? For visualisations, "What am I trying to say?" comes down to which aspect(s) of the data you want to highlight. For example, do you want to focus on total numbers (how many ranked high, how many ranked low, how many in between)? Or do you want to look at a particular criteria and how the pieces scored on it? Do you want to compare video with text and show the relative rates of high & low quality? What is the question you most want this data to answer?

- As a class, combine your findings in a spreadsheet, and then in groups of 3 or 4 find ways to visualise the answer to your question.
- Your first challenge will be to figure out what you want to communicate.
- Your second will be to find the clearest, most compelling way to visualise your results. You can draw the image by hand or in a drawing app, or graph it using Python visualisation libraries or online graph creation packages. It doesn't matter what technology you use here(and yes, pen and paper are technologies!), what matters is your message and how effectively you communicate it.

10

Acknowledgements

So many people have made this book possible. I am immensely grateful for all of the love and support I've received, especially during the endlessly dispiriting lockdowns. Some particular thank yous are in order, but first I have to publicly recognise that none of this would ever have happened without the extraordinary enthusiasm, generosity, and talent of every student I have ever taught, whether in year 10 or year 11, plus the university students who paved the way. You made this possible, and you gave me the insight I needed to make change. You are all amazing!

Big love to all of the people who have supported ADSEI during its birth and toddlerhood. Together you've helped put Data Science Education on a strong footing for the future.

Heartfelt thanks to:

Tia Lowenthal for designing the perfect book cover, and the ADSEI logo, which I still love.

My dedicated proof readers, especially JoAnne King, Gretchen Scott, Rachel Dickenson, Toni Collis, Laura Summers, Laurent Tardif, Javier

Candeira, Andrew McIver, and Jed Wesley Smith. This book is vastly better for your input. All remaining issues are my own. :)

My extraordinary business coach, Dr Toni Collis, for her boundless enthusiasm, endless support, and extraordinary insight.

Sorrell Grogan, Jenny Riesz, Joel Gilmore, and Jim Driscoll for climate science help.

Joshua Heerey and Mark Scholes for introducing me to evidence based physiotherapy, making it possible for me to walk without pain, and for nerding out with me over data and evidence.

Sam Moskwa for many, many coffees, and boundless enthusiasm for ADSEI and my work.

Kathie Mayer, for enthusiasm and, on hearing my pitch saying "oh, so you're raising heretics!" and thus creating the best title ever.

Nick Falkner for keeping me insane but in entertaining ways, and for endless support and good wine.

Nicky Ringland for brainstorming the outline and the final chapter with me, and for always believing in me.

My regular chat room visitors during the endless lockdowns, for keeping me connected and having amazing conversations, especially: Javier, Sumana, Sarah, Nick, Fiona, and Heidi.

Sumana Harihareswara, for amazing conversations and inspiration.

Danny Summerell, and Amanda Hogan for inspiration, enthusiasm, and road testing of ideas.

The ADSEI board, Katrina Falkner, Kathryn Gahan, Penny Hale, Karen Lamb, Ramesh Rajan, maia Sauren, and Robyn Simpson, for supporting this whole mad venture.

Sally-Ann Williams for mentoring me through the startup of ADSEI, and for her unshakeable belief that this would work.

Riley Taylor for encouraging me into action when I was procrastinating by his hospital bed.

The Chatting for Change crew for giving me hope, especially Riley, Caitlyn, Jess, Ben, Josh, Reena, Faris, and Kristina.

My sister, Jane Sykes, for boundless love and support.

My extraordinary twin, Robyn Simpson, for always believing me, for brilliant insights and remarkable conversation, and for love.

My adopted big brother, Mark Stickells, for endless support, patience, encouragement, wisdom, and insight. Also my Pawsey family, for friendship, support, and online lockdown drinks.

The Unbossables, Michele Playfair, Gretchen Scott, Laura Summers, Javier Candeira, and maia Sauren, for virtual coffees, real dinners, and their continuing love and faith in me.

Andrew, Zoe, and Solara, for putting up with the highs and lows of life with someone who needs to change the world or die trying, and for making it clear that this needs doing.

And to Emerald, for patiently chewing up my office while I wrote.

References and Endnotes

1. https://www.abc.net.au/news/2019-06-07/uber-fair-work-ombudsman-investigation-contractor-employee/11189828
2. https://www.wired.com/story/christchurch-shooter-youtube-radicalization-extremism/
3. https://www.washingtonpost.com/technology/2019/10/22/ai-hiring-face-scanning-algorithm-increasingly-decides-whether-you-deserve-job/
4. https://www.theverge.com/2018/4/26/17285058/predictive-policing-predpol-pentagon-ai-racial-bias
5. See any one of the studies described in Utopia for Realists, by Rutger Bregman
6. https://australiainstitute.org.au/wp-content/uploads/2020/12/P245-Company-tax-what-the-evidence-shows.pdf
7. https://www.nngroup.com/articles/ten-usability-heuristics/
8. More detail about this project can be found in Chapter 6
9. https://theconversation.com/the-healing-power-of-data-florence-nightingales-true-legacy-134649
10. https://www.smh.com.au/national/remember-sports-rorts-here-s-why-we-mustn-t-forget-that-shameful-episode-20200925-p55z8x.html

11. https://www.theguardian.com/australia-news/2020/jul/21/sports-rorts-coalition-approved-at-least-six-grants-without-an-application-form-documents-reveal
12. https://www.wired.com/story/buying-giphy-gives-facebook-new-window/
13. https://www.theguardian.com/australia-news/2019/feb/20/more-than-25-million-people-have-opted-out-of-my-health-record
14. https://www.abc.net.au/news/science/2020-06-17/covidsafe-contact-tracing-app-test-documents-rated-poor-iphone/12359250
15. https://www.sbs.com.au/news/scott-morrison-urges-all-australians-to-download-covidsafe-app-says-it-s-the-ticket-to-easing-restrictions
16. For more on this theme, I recommend *Automating Inequality*, by Virginia Eubanks, and *Weapons of Math Destruction*, by Cathy O'Neil
17. https://www.theverge.com/2020/8/17/21372045/uk-a-level-results-algorithm-biased-coronavirus-covid-19-pandemic-university-applications
18. https://www.nature.com/articles/d41586-019-03847-z
19. https://www.nature.com/articles/d41586-019-03847-z
20. https://www.washingtonpost.com/technology/2019/10/22/ai-hiring-face-scanning-algorithm-increasingly-decides-whether-you-deserve-job/
21. The Gendered Brain: The new neuroscience that shatters the myth of the female brain, Gina Rippon, Vintage, 2020.
22. http://prefrontal.org/files/posters/Bennett-Salmon-2009.pdf
23. https://theconversation.com/the-reinhart-rogoff-error-or-how-not-to-excel-at-economics-13646

24. https://www.businessinsider.com.au/graph-shows-georgia-bungling-coronavirus-data-2020-5?r=US&IR=T
25. https://www.abc.net.au/news/2020-05-22/working-from-home-employee-monitoring-software-boom-coronavirus/12258198
26. https://techcommunity.microsoft.com/t5/microsoft-365-blog/microsoft-productivity-score-insights-that-transform-how-work/ba-p/969722
27. https://www.gsb.stanford.edu/insights/good-bad-measuring-worker-output-real-time
28. Weapons of Math Destruction by Cathy O'Neil, Made by Humans by Ellen Broad, and Automating Inequality by Virginia Eubanks
29. https://biometricmirror.com/
30. https://www.finextra.com/newsarticle/31825/blind-woman-sues-cba-over-touchscreen-pos-machines
31. https://fortunly.com/statistics/gig-economy-statistics#gref
32. https://theconversation.com/facebook-is-tilting-the-political-playing-field-more-than-ever-and-its-no-accident-148314
33. https://www.commondreams.org/news/2020/05/26/facebook-ignored-internal-warnings-its-algorithms-were-intensifying-divisiveness
34. https://www.aph.gov.au/About_Parliament/Parliamentary_Departments/Parliamentary_Library/pubs/rp/rp1314/43rdParl
35. for alarming evidence on that front, I recommend Predictably Irrational by Dan Ariely
36. It's important to note that, state-by-state the Australian response was evidence based. The Federal Government was initially keen to follow the USA and the UK on their "herd immunity" path, which directly contradicted the available evidence and has resulted in hundreds of thousands of unnecessary deaths.

37. https://sustainable.unimelb.edu.au/news/what-are-the-full-economic-costs-to-australia-from-climate-change
38. https://doi.org/10.1007/s40279-019-01092-y
39. https://fivethirtyeight.com/features/surgery-is-one-hell-of-a-placebo/
40. https://medicine.uq.edu.au/article/2017/06/explaining-vaginal-mesh-controversy
41. https://www.vox.com/the-big-idea/2017/12/28/16823266/medical-treatments-evidence-based-expensive-cost-stents
42. https://www.ncbi.nlm.nih.gov/pmc/articles/PMC4800017/
43. https://www.hopkinsmedicine.org/health/wellness-and-prevention/is-taking-aspirin-good-for-your-heart
44. https://www.ncbi.nlm.nih.gov/pmc/articles/PMC4417373/
45. Bergs J, et al. Systematic review and meta-analysis of the effect of the World Health Organization surgical safety checklist on postoperative complications. BJS 2014;101:150–158
46. https://www.fda.gov/consumers/consumer-updates/grapefruit-juice-and-some-drugs-dont-mix
47. https://www.immdsreview.org.uk/news.html
48. https://www.esrl.noaa.gov/gmd/ccgg/trends/graph.html
49. https://climate.nasa.gov/news/2535/satellite-data-confirm-annual-carbon-dioxide-minimum-above-400-ppm/
50. https://www.jstor.org/stable/4314542?seq=1
51. https://www.euro.who.int/__data/assets/pdf_file/0006/189051/Health-effects-of-particulate-matter-final-Eng.pdf
52. https://www.lazard.com/perspective/levelized-cost-of-energy-and-levelized-cost-of-storage-2020/
53. https://www.nature.com/articles/s41598-020-66275-4
54. https://climate.nasa.gov/effects/
55. https://www.uts.edu.au/sites/default/files/article/downloads/ISF_100%25_Australian_Renewable_Energy_Report.pdf

56. https://www.abc.net.au/news/2020-09-16/renewable-energy-agencies-allowed-invest-low-emission-technology/12670760
57. https://www.youtube.com/watch?v=Avxm7JYjk8M
58. https://sustainable.unimelb.edu.au/__data/assets/pdf_file/0012/2756874/MSSI-IssuesPaper-10_Last-Resort-Housing_2017_0.pdf
59. https://www.imf.org/en/Publications/Staff-Discussion-Notes/Issues/2016/12/31/Causes-and-Consequences-of-Income-Inequality-A-Global-Perspective-42986
60. https://www.tai.org.au/sites/default/files/P709%20Tax%20and%20wellbeing%20%5BWEB%5D.pdf
61. https://www.acoss.org.au/wp-content/uploads/2018/02/010218-Cashless-Debit-Card-Briefing-Note_ACOSS.pdf
62. https://www.vox.com/future-perfect/2020/2/19/21112570/universal-basic-income-ubi-map
63. https://www.unaa.org.au/2017/06/28/unesco-study-reveals-correlation-between-poverty-and-education/
64. https://theconversation.com/atars-you-may-as-well-use-postcodes-for-university-admissions-19154
65. https://www.dese.gov.au/download/1307/review-funding-schooling-final-report-december-2011/1280/document/pdf
66. https://www.acer.org/au/discover/article/measuring-the-extent-of-out-of-field-teaching
67. https://www.gse.harvard.edu/news/uk/16/09/intrinsically-motivated.
68. https://doi.org/10.1002/ajs4.142
69. https://amsi.org.au/wp-content/uploads/2019/01/researchreport4-maths_anxiety_students_and_teachers.pdf
70. https://www.swinburneonline.edu.au/online-courses/education/bachelor-education-primary?gclid=Cj0KCQiAwMP9BRCzARIsAPWTJ_Hh4fMpdTmSiVE7hlhPwSTLRXVabJx8teBXr

kLjEx_WCqEcoD4ApbsaAuTmEALw_wcB&gclsrc=aw.ds (accessed 16 November 2020)
71. https://handbook.une.edu.au/courses/2021/BEDK6 (accessed 16 November 2020)
72. https://www.open.edu.au/degrees/bachelor-of-education-primary-education-curtin-university-cur-bed-deg (accessed 16 November 2020)
73. https://helix.northwestern.edu/article/thalidomide-tragedy-lessons-drug-safety-and-regulation
74. https://www.wired.com/story/the-teeny-tiny-scientific-screwup-that-helped-covid-kill/
75. https://scholarspace.manoa.hawaii.edu/bitstream/10125/22778/1/vol63n4-601-616.pdf
76. https://linkinghub.elsevier.com/retrieve/pii/S0163638300000321
77. https://www.youtube.com/watch?v=nWu44AqF0iI
78. https://www.theatlantic.com/family/archive/2018/06/marshmallow-test/561779/
79. https://timssandpirls.bc.edu/timss2019/encyclopedia/index.html
80. https://www.covid19data.com.au/victoria
81. https://www.uac.edu.au/assets/documents/atar/usefulness-of-the-atar-report.pdf
82. https://www.monash.edu/students/handbooks/outcomes
83. https://www.adelaide.edu.au/learning/resources-for-educators/graduate-attributes
84. Names have been changed to protect the infuriating.
85. PISA stands for Program for International Student Assessment. It's an OECD benchmarking test for 15 years olds around the world. "PISA is the OECD's Programme for International Student Assessment. PISA measures 15-year-olds' ability to use their read-

ing, mathematics and science knowledge and skills to meet real-life challenges."
86. https://www.abc.net.au/news/2016-08-09/abs-website-inaccessible-on-census-night/7711652
87. https://www.abc.net.au/news/2016-08-01/census-2016-why-are-people-worried-about-the-census/7678198
88. I was *not* raised to be a heretic. I am self-taught, and occasionally lapse into unthinking obedience.
89. https://www.ncsehe.edu.au/wp-content/uploads/2014/12/Li-and-Dockery-Schools-SES-Final.pdf
90. https://www.monash.edu/__data/assets/pdf_file/0006/133971/emmaline-bexley-presentation.pdf
91. https://www.theage.com.au/education/the-epidemic-australia-is-failing-to-control-20201229-p56qq3.html
92. https://onlinelibrary.wiley.com/doi/abs/10.5694/mja13.10103
93. https://www.scientificamerican.com/article/why-is-the-sky-blue/
94. https://www.mayoclinic.org/diseases-conditions/hyponatremia/symptoms-causes/syc-20373711
95. https://www.npr.org/templates/story/story.php?storyId=106268439
96. https://www.medicalnewstoday.com/articles/255712#The-BMI-formula
97. https://elemental.medium.com/the-bizarre-and-racist-history-of-the-bmi-7d8dc2aa33bb
98. https://www.medicalnewstoday.com/articles/265215#BMI-exaggerates-thinness-in-short-people-and-fatness-in-tall-people
99. https://www.euro.who.int/en/health-topics/disease-prevention/nutrition/a-healthy-lifestyle/body-mass-index-bmi
100. https://locusmag.com/2021/05/cory-doctorow-qualia/

101. https://www.wired.com/story/the-teeny-tiny-scientific-screwup-that-helped-covid-kill/
102. https://researcher.watson.ibm.com/researcher/view_group.php?id=4933
103. https://www.abc.net.au/news/2020-10-06/excel-error-which-led-to-uk-coronavirus-test-debacle/12734754
104. https://commons.wikimedia.org/wiki/File:Victorian-Senate-ballot-paper-2016.svg
105. https://www.aec.gov.au/voting/how_to_vote/voting_senate.htm
106. https://docs.google.com/document/d/1Xfr2YSsF-mORn5Jh1LdSe2ASq46Hw7nmL_jmaqCdb2IU/edit?usp=sharing
107. https://stem.pawsey.org.au/
108. https://en.wikipedia.org/wiki/Anscombe%27s_quartet
109. http://www.cleanenergyregulator.gov.au/RET/Forms-and-resources/Postcode-data-for-small-scale-installations
110. https://www.tylervigen.com/spurious-correlations
111. https://www.digitaltechnologieshub.edu.au/teachers/lesson-ideas/integrating-digital-technologies/humpback-whales-what-the-data-reveals
112. https://www.theage.com.au/national/covid-19-global-vaccine-tracker-and-data-centre-20210128-p56xht.html
113. https://teaching.cornell.edu/teaching-resources/engaging-students/problem-based-learning

www.ingramcontent.com/pod-product-compliance
Lightning Source LLC
Chambersburg PA
CBHW070252010526
44107CB00056B/2430